CLOCK AND COMPASS

Clock & Compass

HOW JOHN BYRON PLATO GAVE FARMERS A REAL ADDRESS

Mark Monmonier

University of Iowa Press,
Iowa City

University of Iowa Press, Iowa City 52242

www.uipress.uiowa.edu
Printed in the United States of America

Design by Ashley Muehlbauer

Printed on acid-free paper

Library of Congress Cataloging-in-Publication Data

Names: Monmonier, Mark S., author.
Title: Clock and Compass: How John Byron Plato Gave Farmers a Real Address / Mark Monmonier. Description: Iowa City: University of Iowa Press, 2022. Identifiers: LCCN 2021030561 (print) | LCCN 2021030562 (ebook) |
ISBN 9781609388218 (paperback) | ISBN 9781609388225 (ebook)
Subjects: LCSH: Plato, John Byron, 1876–1966. | Compass system (Cartography) | Rural geography—Maps—United States—History. | Farms—Location—United States. | Cartographers—United States—Biography. Classification: LCC GA407.P53 M66 2022 (print) | LCC GA407.P53 (ebook) | DDC 526/.64—dc23
LC record available at https://lccn.loc.gov/2021030561
LC ebook record available at https://lccn.loc.gov/2021030562

For my grandsons, Camden and Cooper

CONTENTS

PREFACE

Clock and Compass is an outgrowth of my book *Patents and Cartographic Inventions*, written to offer (as its subtitle promised) "a new perspective for map history." By calling attention to the several hundred map-related inventions among the several million clever ideas vetted by the US Patent Office during its first two centuries, I introduced academic historians of cartography and map enthusiasts in general to a broad range of ingenious but little-known strategies for pinpointing places, navigating highways, folding maps, projecting worldviews, manufacturing globes, and exploring the promise of electronic circuitry. Few of these patents received wide recognition—Buckminster Fuller's iconic Dymaxion projection and some clever starburst folding schemes are prominent exceptions. Most merely demonstrated that not all clever ideas are worth marketing. The patents system certifies originality but cannot guarantee sales, profit, or even an uncertain launch.

Among the various inventions featured in *Patents*, one stands out: John Byron Plato's "Clock System," which uses distance and direction from a nearby business center to give farmhouses an address as specific and workable as the house numbers and street names used in cities. Devised several decades before utility companies, town governments, and E-911 offices extended street-numbering into the countryside, the Clock System depended upon both a map and a directory, which also accommodated the paid advertising that supported compilation and distribution. At a time when most rural residents lacked telephone service, it played a role similar to the phonebook, which might have contributed to its demise. At the nexus between residents, advertisers, and the directory publisher, the Clock System occupies a small but not insignificant place in the history of wayfinding and telecommunication technologies that evolved into the satnav and the internet. Plato's invention fully merited its featured treatment in *Patents'* second chapter, which focused on geographic location.

I first learned of the Clock System a few decades ago in the 1980s, though I cannot be certain. I vaguely recall browsing the map library at Syracuse University, where I am on the geography faculty, and finding a Compass System map and directory for Oswego County, New York, which lies between Onondaga County, where I live, and Lake Ontario. I photocopied a few key pages and stuck them in a "for-later-use" folder, in a file drawer where I stockpiled ideas for new projects. Note that I said Compass System, not Clock System, because the Oswego map, copyrighted in 1938, relied on Plato's basic framework (business center, distance, and direction) but was published by a firm that revived Plato's game plan five years after he left town at the onset of the Great Depression.

That few Clock System and Compass System maps have survived in libraries, public map collections, and personal caches of antique road maps is both understandable and puzzling. Understandable because ephemera, such as phonebooks and flyers, are often deemed too common, too fleeting, and thus too insignificant to merit systematic archiving and cataloging. And puzzling because their detailed and fascinating pictures of past landscapes are of obvious value to local historians and collectors of road maps and other Americana once important to everyday citizens. More troubling is the apparent fact that almost all the maps Plato filed with his copyright applications have gone missing. The Library of Congress, which normally would receive one of the two "deposit copies," lists only his "Clock System Map of Onondaga County" in its online catalog. The map, never registered for copyright when it was published in 1927, was donated to the Library in 1992.

Clock and Compass attempts to construct as revealing a biography of Plato as available evidence might allow, while weaving in key details about his invention and its impacts as well as his business model and collaborators. The framework is largely chronological, starting with a concise description of the problem Plato sought to solve: a problem rooted in rural sociology as well as cartography. My research strategy for probing his solution is both archival and analytical. Newspaper stories retrieved through searchable databases supplement archived census materials, city directories, and other genealogical tools collectively known as Big Microdata. Additional evidence includes the temporal and geographic patterns of mapping activity as documented by Plato's copyright registrations, and the form and content of the maps themselves. As with my previous books, I rely heavily on maps and diagrams to tell the story. Some are facsimiles, chosen to illustrate what Plato was up to. Others are my own designs, intended to summarize geographic relationships; discussion of their development will, I hope, help the reader appreciate that maps are not only part of the story but part of the storytelling.

As an academic sleuth, I sought to find and connect as many relevant dots as needed to craft a coherent story about what Plato did, why he did it, and with what impact. I cannot say, "as many relevant dots *as possible*" because a researcher can always initiate one more search or look under one more rock. As a map historian, I must rely on interpolation and extrapolation, and use "apparently" to flag iffy explanations and interpretations. Plato might have helped by donating his business records to a historical society, or by marrying young and siring observant offspring, or at least by being born into a tribe of long-lived kinfolk committed to preserving his memory—but he didn't. That said, he enjoyed writing and talking to news reporters, and left a host of facts, which I could usually (but not always) triangulate. Not everything is knowable, but by and large I think I got it right.

MARK MONMONIER
DeWitt, New York

CLOCK AND COMPASS

1 NO REAL ADDRESS

Relative isolation put rural residents near the forefront of the nation's rapid shift in personal transportation from steel rails to paved roads. In 1916 the United States railway network reached its peak of 254,037 route miles.[1] Although the railnet still registered a robust 249,052 miles by 1930, these oft-cited estimates are deceptive: steam railroads had made long-distance travel ever more practicable, at least between stations with passenger service. Streetcar systems extended this accessibility within cities large and small, and electric interurban lines put some of the adjacent countryside, and many farmers, within reach.[2] Nonetheless, these fixed-route conveniences did not serve everyone and were not to last. The principal disrupter was the motor car.

Before the 1920s became the age of automobile design, improved reliability and lower prices had made the 1910s an age of personal motorized transport. Industrious carriage makers and machinists became automobile manufacturers, and car dealerships found willing buyers in both urban and rural settings.[3] Some farm-machinery dealers discovered car sales a profitable sideline as well-off or debt-tolerant farmers recognized the automobile as an affordable supplement to their more utilitarian pick-up truck. In cities and small towns, affluent homeowners delighted in a new use for their carriage house or backyard stable. Great wealth was not a prerequisite for city residents dissatisfied with the dubious convenience of an electric streetcar network focused on downtown. Although only 2 percent of American households owned a car in 1910, automobility increased markedly over the next two decades. The federal census did not count households with cars, but a *New York Times* statistician estimated per-household car-ownership rates of 28 and 59 percent for 1920 and 1930, respectively.[4] Outside medium and large cities, where the trolley remained a convenient alternative, proportionately more households owned cars.

Fueling the transition to car ownership was the petroleum industry, which followed the discovery of oil in Titusville, Pennsylvania, in 1859. Production initially focused on kerosene, used principally for illumination as a substitute for whale oil. Although the earliest motor cars ran on natural gas, which was difficult to produce and store, improved internal-combustion engines created a profitable use for gasoline, an overly volatile and largely useless byproduct of kerosene production.[5] Discovery of oil fields elsewhere in Appalachia and in Kansas and Oklahoma fostered the creation of John D. Rockefeller's Standard Oil monopoly, broken up by a 1911 Supreme Court decision. Successor companies like Esso and Socony developed networks of local gasoline and motor-oil dealers, essential to long-distance motoring.[6]

Although a dirt road still ran past the typical farm, an expanding network of paved highways increased the rural traveler's range and aspirations. In the 1870s bicyclists eager for solid, stable pavement free of mud, stones, and dust initiated the Good Roads Movement, embodied by the League of American Wheelmen, founded in 1880.[7] The League lobbied state governments for smooth roads, paved with macadam, another recent innovation, and started its own magazine, *Good Roads*, in 1892.[8] The movement gained further support from the Grange Movement and the US Post Office, which was experimenting with Rural Free Delivery service in the 1890s.

In the 1910s turn-by-turn instructions in guidebooks, sometimes reinforced with simple maps and photographs, helped motorists navigate the irregular and poorly marked routes connecting cities and larger towns.[9] In the mid-1920s the American Association of State Highway Officials provided a more stable framework for small- and intermediate-scale road maps by establishing a nationwide numbering scheme with standardized signage.[10] The free road map emerged in the late 1920s, when low-cost printing and economies of mass production at Rand McNally, General Drafting, and H.M. Gousha intersected the competitive needs of petroleum companies, such as Esso and Texaco, with supply chains supporting thousands of franchised or company-owned gas stations.[11] Up-to-date disposable maps encouraged motorists to find a route and just follow the signs. However helpful these navigation aids, uncertainty arose when the destination was off the numbered network.

In the era before farms acquired house numbers, rural travelers venturing outside their immediate neighborhood or beyond the network of numbered highways often relied on hand-drawn maps or verbal directions. Locally prominent visible landmarks were useful reference points, as when a traveler was told, "Go about three miles until you see a large red barn, and then bear left." Countable guidance like "fourth house on the right" underscored the need to show individual structures to reinforce a reliable representation of noteworthy bends and straight stretches. Today's electronic navigation systems recognize the value of locally distinctive landmarks, as when a motorist's GPS says, "Turn left at the gas station," or "turn right at the second set of traffic lights."

Farmers sufficiently prosperous to own one might consult the family's county land-ownership atlas. In the latter half of the nineteenth century a team of surveyors would descend on a county, check names of householders, and measure distances along roads with a wheelbarrow odometer, also called a surveyor's perambulator (fig. 1).[12] They consulted plat maps and other data readily available at the county courthouse, and compiled maps for oversize atlas pages (roughly 17 × 14 inches). Because the surveyor-canvassers also took orders for the finished atlas—buyers eager to see their house and name on the map had to prepay—publishers of these so-called subscription atlases had no need to front the costs of engraving and printing.[13] As a further appeal to vanity, for an additional fee the atlas could include an engraved sketch of the buyer's home embellished with mature shade trees and meticulously trimmed ornamental plantings. The lucrative county atlas business, along with spinoffs such as county wall maps and state atlases, accounted for four of the eighteen chapters in muckraker Bates Harrington's 1890 book *How 'Tis Done: A Thorough Ventilation of the Numerous Schemes Conducted by Wandering Canvassers, Together with the Various Advertising Dodges for the Swindling of the Public.*[14]

Despite the hype and high-pressure salesmanship, these county-format atlases have been praised as generally reliable representations of the late-nineteenth-century American landscape, particularly useful for counties where original ownership records were lost in a fire or flood.[15] Coverage was notably thorough in rural parts of the Central and Mid-Atlantic states, where local landowners able to afford them were relatively numerous. Clara Le Gear, who compiled a cartobibliography for the Library of Congress, identified county atlases published in the 1860s and 1870s

FIGURE 1. Wheelbarrow odometer, a push-cart device for measuring distance along roads. From Bates Harrington, *How 'Tis Done: A Thorough Ventilation of the Numerous Schemes Conducted by Wandering Canvassers; Together with the Various Advertising Dodges for the Swindling of the Public* (Syracuse, NY: W. I. Pattison, 1890), 17.

covering fifty of the fifty-five New York counties north of New York City.[16] By 1900 an atlas had been published for most counties in the agricultural states of the Midwest. These atlases enrich the reference collections of county historical societies and genealogical departments of public libraries fortunate to own one.

Subscription atlases were designed to flatter as well as inform. Printed at scales of approximately 1:31,680 (two inches to the mile), their maps were sufficiently detailed to show individual homes and residents' names, as well as the orientation of a farmstead's rectangular footprint.[17] As illustrated in figure 2, from an atlas published in 1874, the engraver often had a preference for jerky straight-line segments, no doubt easier to inscribe than the more realistic smooth curves that later mapmakers could delineate efficiently with pen and ink. Although this atlas made it easy to pinpoint some destinations, it was of limited use for in-vehicle navigation by a 1920s' motorist because many of the residences shown had disappeared. And why risk damaging a would-be family heirloom on a local road trip? Putting farmers on the map did little to help them with wayfinding.

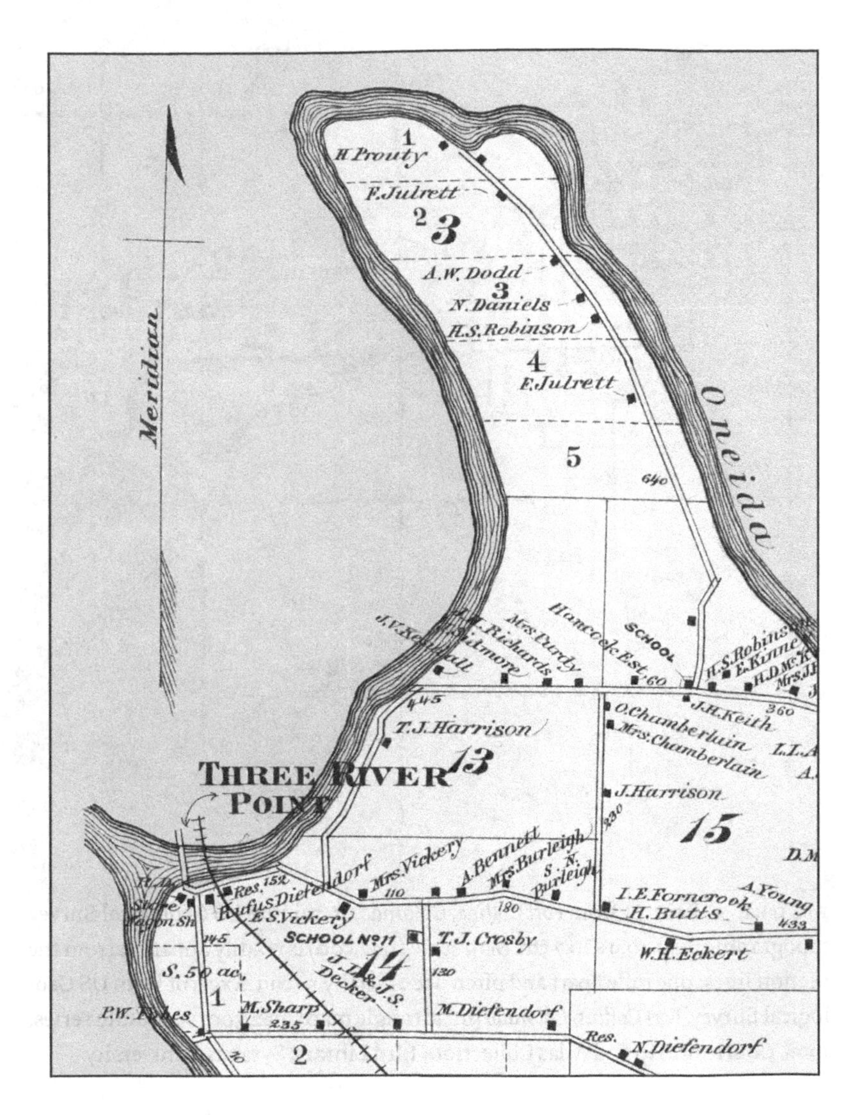

FIGURE 2. An 1874 county atlas provides a detailed rendering of a neighborhood in the northwestern corner of the Town of Clay, in Onondaga County, New York. Excerpt from the northwestern corner of the map for the Town of Clay, in *Sweet's New Atlas of Onondaga Co., New York: From Recent and Actual Surveys and Records under the Superintendence of Homer D. L. Sweet* (New York: Walker Bros. & Co., 1874), 16–17. Courtesy Map and Atlas Collection, Bird Library, Syracuse University.

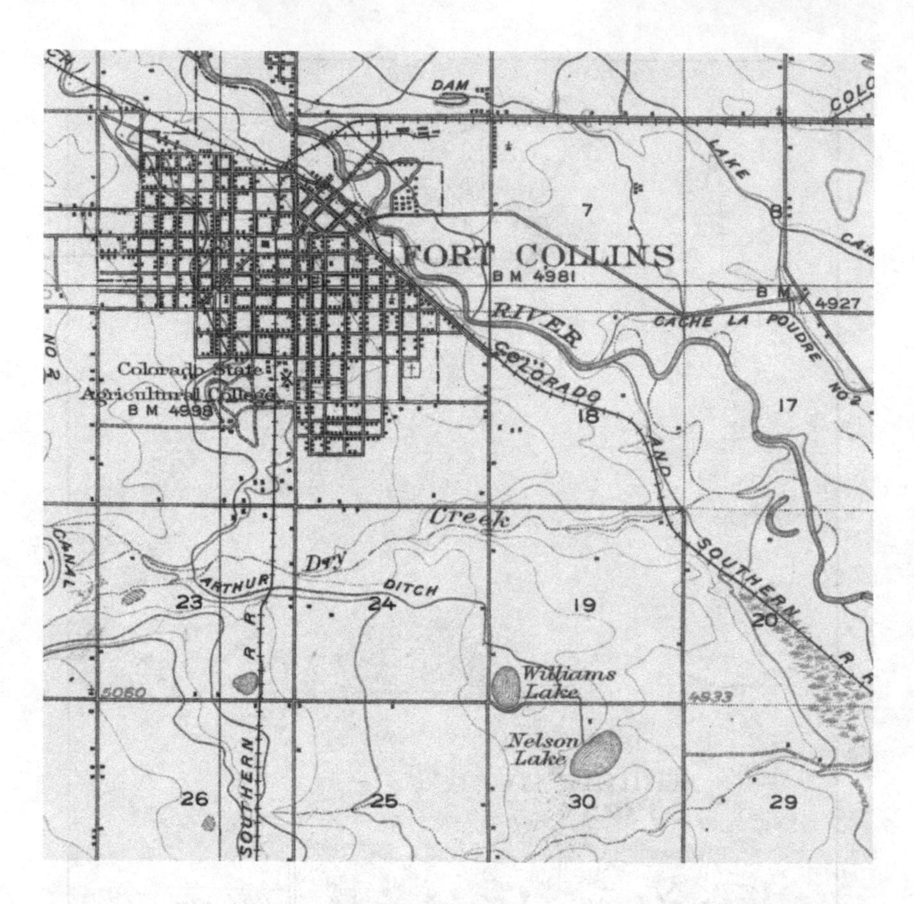

FIGURE 3. Excerpt from *Fort Collins, Colorado*, 15-minute US Geological Survey topographic map. 10.9 × 10.7 cm. Map scale (1:62,500) is readily apparent from the section lines, one mile apart and often occupied by a road. Excerpt from US Geological Survey, *Fort Collins, Colorado* (quadrangle map), 1:62,500, 15-minute series, 1906. Courtesy Map and Atlas Collection, Bird Library, Syracuse University.

Another cartographic product suitably detailed to show individual farmsteads was the US Geological Survey (USGS) topographic map, typically published at a scale of 1:62,500 (roughly one inch to the mile). Established in 1879, the USGS initiated an ambitious program of quadrangle-format mapping, but coverage was a slow build.[18] Although the typical quadrangle encompassed 15 degrees of latitude and 15 degrees of longitude, some of the early USGS quadrangle maps covered four times the area (30 degrees of latitude and 30 degrees of longitude) on the same size sheet (roughly 17 × 21 inches) at half the scale, 1:125,000 (roughly one inch to two miles).

The excerpt in figure 3, from a 15-minute quadrangle map of Fort Collins, Colorado, published in 1906, includes several basic feature categories. Though not apparent on this black-and-white version, color coding distinguishes natural features from human imprints like infrastructure and feature names. Lakes, ponds, reservoirs, streams, rivers, and irrigation canals—collectively known as hydrography—are blue; elevation contours, spot elevations (at selected road intersections), and dams are brown; and roads, railways, structures, and other forms of "public culture" are black.

Many feature types are apparent from their distinctive shape or accompanying labels: mostly straight double-line symbols represent roads, heavier single lines with small, evenly spaced cross-ticks portray railroads, and alternating long and short dashes delineate the municipal boundary. The dominant water feature is the Cache La Poudre River, flowing from northwest to southeast and receiving drainage from perennial streams (thin continuous lines) as well as seasonal streams (long dashes separated by three dots) such as Dry Creek, about a mile south of the city line. Irrigation canals, such as Arthur Ditch (south of Dry Creek), reflect a semiarid climate in which farmers depend on an artificial water supply. Italic type for water features reflects a well-established cartographic convention.

Elevation contours are everywhere. With a vertical interval of 20 feet, these brown lines are widely spaced where the land slopes gently toward the river and closer together where the land pitches more steeply toward a tributary. Within and immediately adjacent to the city boundary are numerous small black rectangles representing structures. Aside from the few rectangles with a cross indicating a church, the map makes no distinction among residential, commercial, industrial, and governmental uses. Along city blocks with closely spaced structures the rectangles coalesce. Outside the city the map shows farmsteads but usually omits barns and other outbuildings. The smooth curves of the Colorado and Southern Railway, which follows the meandering river on higher ground, contrast with the two comparatively straight tracks that converge on the city from the east and south. Waterways and railroads are named, but the area's more numerous roads remain anonymous—a clear limitation for anyone wanting to use a USGS topographic map as a wayfinding aid.

Even so, a motorist might find some solace in the general pattern of roads running north–south or east–west and spaced a mile apart, as if on a grid. After all, an odometer could be useful in following verbal directions gleaned from the map; for instance, "from the center of town go five miles east; then turn right and go three miles south; then turn left, go about a quarter mile, and look for the white house on the right with the name 'G. Jones' on the gatepost." This instruction might work in many parts of the Fort Collins area, but not all sections of the rectangular grid have a double-line road symbol and not all roads follow a cardinal direction. Although the grid could be a basis for street naming and assigning addresses, this was not the practice in the early twentieth century.

The area's rectilinear road pattern and similar grids in many other states reflect a recipe for subdividing land known as the Public Land Survey System (PLSS).[19] Created after the Revolutionary War when the new nation began acquiring territory west of the original thirteen colonies, the PLSS was a systematic strategy for describing tracts of land that the government would eventually sell, allocate for worthy purposes like education and railroads, or give to settlers who agreed to occupy and put to productive use an allotment of 160 acres. The system was implemented piecemeal, with individual survey districts covering all or part of a state, or perhaps multiple states. Each survey district was framed by a "principal meridian" and a parallel called a "base line," which intersected at an "initial point." As described in the lower part of figure 4, a survey district was divided into square "townships," six miles on a side and organized into a grid of rows (also called townships) and columns called ranges. Because township rows are numbered north or south of the base line and ranges are numbered east or west of the principal meridian, a township's location can be referenced by its pair of township and range numbers. Examples labeled (T.2S, R.3W) and (T.2N, R.3E) illustrate this strategy.

PLSS locations are hierarchical, which makes the system useful for subdividing land and setting up local governments but too complicated for giving directions, assigning addresses, or general wayfinding. Each township is divided into thirty-six one-square-mile "sections," measuring one mile on each side and numbered in the meandering zigzag scheme shown in the upper right part of figure 4. In turn, each section encompasses 640 acres, which can be subdivided into four 160-acre quarter sections. Considered large enough to support a family, the quarter section was the

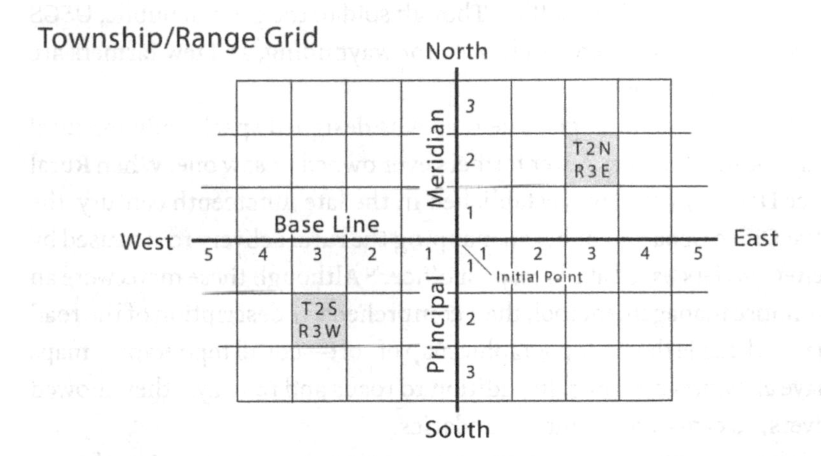

FIGURE 4. Key elements of the Public Land Survey System include township and range numbers referenced to the base line and principal meridian (lower) and the zigzag numbering of square-mile sections within each thirty-six square-mile township (upper right). Diagram in the upper left locates the two townships covering the area in Figure 3. Compiled by the author.

typical grant to homesteaders. Its locational reference begins on the left with the fractional part (e.g., SE ¼, Sec. 23, T.2N, R.3E). A quarter section can be subdivided further into quarter-quarter sections of 40 acres.

On USGS quadrangle maps for a PLSS area, township and range numbers can be found in the comparatively vacant part of the map sheet that mapmakers have dubbed the "collar," with township numbers down the left and the right, and range numbers along the top and the bottom. Within the map proper, sections are numbered according to the prescribed zigzag scheme (fig. 4, upper right). For example, in figure 3 the label for Sec. 23, T.7N, R.69W is slightly below and to the left of the A in "ARTHUR DITCH" (in the lower-left portion of the excerpt), and Sec. 24 is immediately to its right. Section boundaries not apparent in the road system are portrayed by thin vertical and horizontal lines (right side of fig. 3), distinctly weaker than the double-line road symbols. That the excerpt straddles parts of two townships is apparent because immediately to the right of Sec. 24, T.7N, R.69W is Sec. 19, T.7N, R.68W, with Williams Lake in its lower-left corner. The vertical boundary between the two sections runs northward through the "t" in Fort Collins. Though sold to the general public, USGS topographic maps were rarely used for wayfinding, and few farmers are likely to have owned one.

Another kind of large-scale map was designed specifically for rural wayfinding, but even fewer farmers ever owned or saw one. When Rural Free Delivery (RFD) was established in the late nineteenth century, the Post Office Department began mapping the rural delivery routes used by letter carriers based at a local post office.[20] Although these maps were an in-house management tool, their comprehensive description of the road network made them a topographic map of sort—not all topographic maps have elevation contours. In addition to roads and railways, they showed rivers, streams, and county boundaries.

The Post Office had been using maps since its inception. Local postmasters obviously needed to know their station's immediate neighbors. Moreover, because postal employees or contractors who carried mail from station to station followed prescribed routes with measured distances, detailed maps were essential for planning arrivals and departures as well as for setting compensation. In 1837 an in-house Division of Topography began replacing outside mapmakers.[21] Its key product was the intercity postal-route map, which encompassed entire states or parts of adjoining states. Intercity maps included county boundaries and, well before the century's end, railroads with a mail contract.[22] Letters traveling a hundred miles or more moved mostly by train.

RFD service began as an experiment with three routes in West Virginia in 1896, when the Post Office surveyed forty-three additional routes in twenty-nine states.[23] By 1899 RFD service had increased to 412 routes, spread across forty states and one territory. That year the Post Office Department established the first countywide mail service in Carroll County, Maryland, roughly forty-five miles northwest of Washington, DC. Other maps, narrower in scope, described routes emanating from a single local post office. The number of individual routes rose steadily, reaching 43,866 in 1915, when it leveled off.[24] Residents could propose a route that was less than twenty-five miles long, on a gravel or macadam surface, with at least a hundred families within a quarter mile of the route; delivery was only to boxes along the designated route, and the application had to include a map.[25] The Post Office regulated the size and shape of the rural mailbox, with its distinctive attached flag that could be raised to tell the carrier to pick up outgoing mail.[26]

Rural-delivery maps were essential because local postmasters needed to know that all designated mail recipients were on one of their routes, that the local system of routes was efficient and duplication was minimal, and that workload was more or less equitable or carriers were appropriately compensated. Intra-government cooperation inspired the Post Office to share these maps with other federal agencies, most notably the Army (especially for Corps of Engineers projects and as predecessor of the Air Force) and the Department of Agriculture (for marketing and soil survey activities). After the Department began selling maps to the public in 1908, postal cartography informed many of Rand McNally's various local and regional maps and atlases. The biggest single buyer was the Fuller Brush Co., which used rural-delivery maps to facilitate door-to-door sales.[27]

Rural-delivery maps told the mail carrier where to go and in what sequence. On the 1911 edition of the Carroll County rural-delivery map (fig. 5), bold lines and arrows show each carrier's prescribed movement along a route identified by a letter, a number, or both. Neighboring routes intersect with minimal overlap. Double-dashed lines represent unimproved roads, carefully avoided, and surnames identify mail recipients at turning points. In an elaborate countywide dance, carriers anchored to a home post office ventured forth to navigate a complex path with whatever zigzags and reversals were needed to reach all residents.

The excerpt in figure 5 focuses on route K, out of Keymar, Maryland, but also shows part of route K2. Both routes begin at the Keymar post office, marked by a dot-within-a-circle symbol just south of the railroad track, and run a quarter mile south toward the Dayhoff residence, where

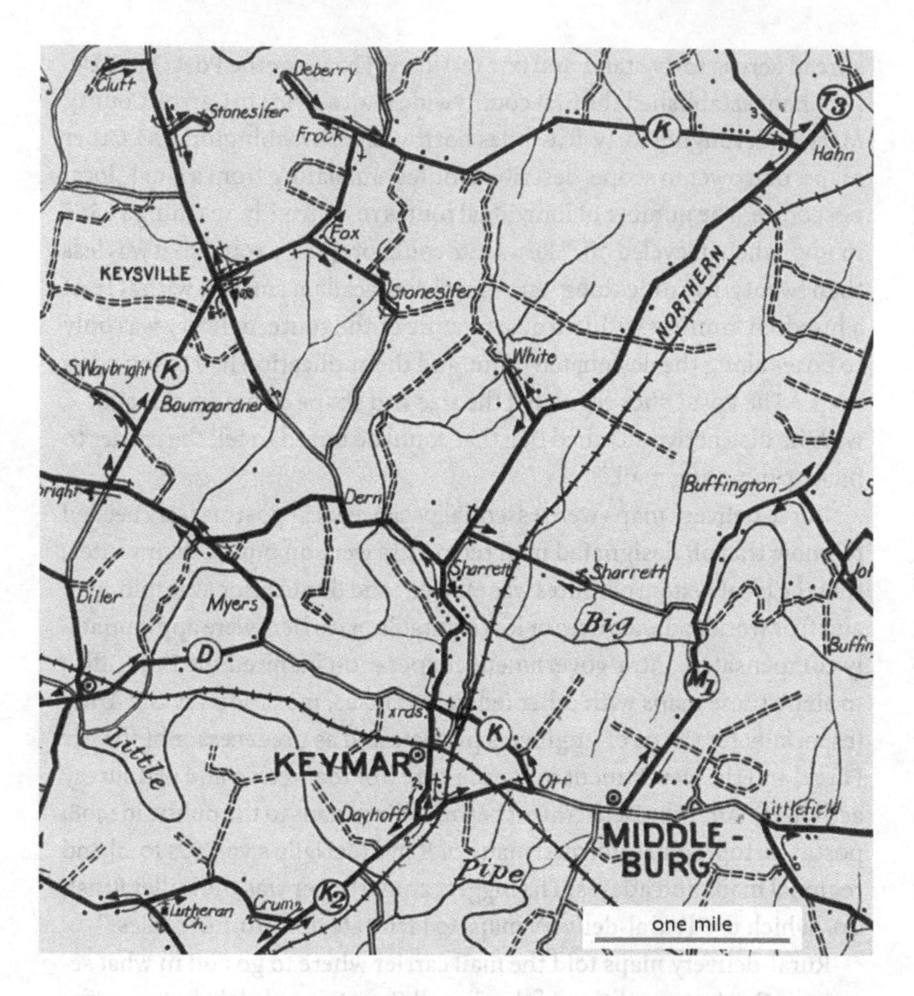

FIGURE 5. Excerpt from 1911 rural-delivery route map of Carroll County, Maryland, describes route K, based at the Keymar post office, 10.8 × 12.0 cm. Excerpt from US Post Office Department, *Map of Carroll County, Md., Showing Rural Delivery Service* (Washington, DC: Post Office Department, 1911). Digital image from the Library of Congress, https://www.loc.gov/item/2012585334/.

route K turns east. Tiny arrows describe a generally clockwise path that reaches the Clutt household (at the upper left) and the Hahn residence (in the upper right) before turning south toward the starting point. Along the way the map prescribes numerous side trips, with out-and-back movements reaching residences off the general circuit. Small dots akin to the structures on a USGS map mark numerous unidentified homes along the route. The carrier learned the recipients' names, often inscribed on their mailbox along with a box number.

For a substitute or new letter carrier learning the route, the rural-delivery map was indeed a wayfinding map. For nonemployees, not so much. Not sold locally, the maps were of limited use unless a local postal worker who knew the route marked the specific destination. (Or unless the visitor consulted the local property map at the courthouse.) Because the locations of rural residences were very much a matter of local knowledge, an outsider would have faced the daunting task of driving the route looking for the right mailbox or asking neighbors along the way.

Wayfinding was far easier in cities, where street names and house numbers gave everyone an address useful for more than delivering mail. And by century's end most large or growing cities inspired at least one enterprising cartographer to create a detailed map that not only named all the streets but also included an index that pointed the user to a specific part of the map.[28] Expanding cities, such as turn-of-the-century Denver, were an attractive cartographic market because visitors and new residents needed a street guide and urban growth soon made older editions obsolete. Advertising blocks on folded editions sold at newsstands, bookstores, drug stores, and other high-traffic retailers helped the map publisher keep the price low and capture market share. Volume sales to banks and real estate dealers made the map a promotional tool useful for attracting customers with a free folded street guide bearing the firm's name, location, and other specifics on an outside panel.[29]

Despite its straightforward content, the indexed street map challenged its creator's ingenuity.[30] As a commercial product with a thin profit margin, the map's single-sheet rectangular format constrained map scale, which in turn could have a profound effect on usefulness and aesthetics. Additional tradeoffs arose in the selection of a printing method, use of color, sheet size, folding scheme, typeface, and lettering size, as well as the allocation of panels for advertising, local history, front and back covers, and the street index. A key decision was the map's geographic scope—how large an area to cover—and which more densely settled areas warranted a detail inset or perhaps their own, relatively large portion of the printed sheet, which raised the further issue of how much redundant coverage to allow. Printing on both sides of the paper allows for a smaller, less wieldy sheet but

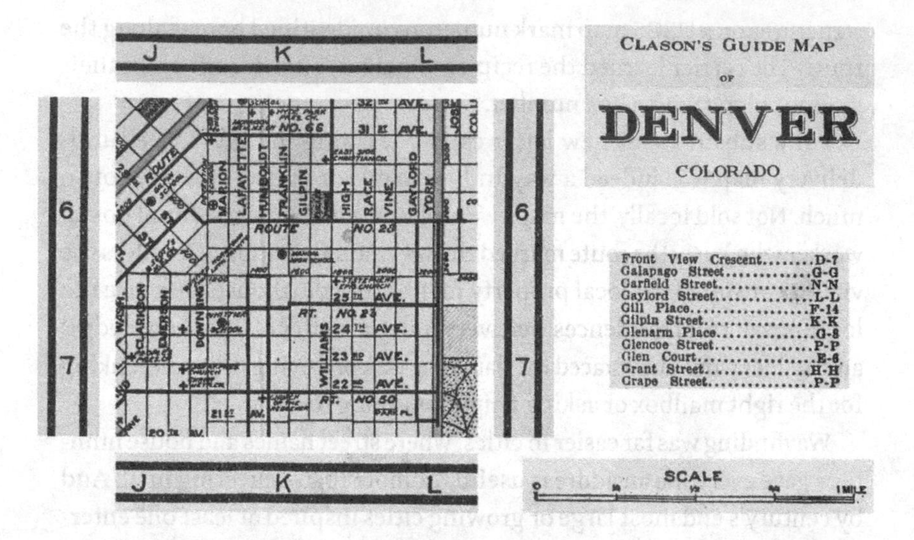

FIGURE 6. Graphic excerpts describing the grid-reference and street-index schemes on *Clason's Guide Map of Denver, Colorado* (1917); street map excerpt is 4.8 × 4.4 cm. Excerpts from *Clason's Guide Map of Denver, Colorado* (Denver: Clason Map Co., 1917). Digital image from Western History and Genealogy Dept., Denver Public Library, https://digital.denverlibrary.org/digital/collection/p16079coll39/id/811.

invites questions about whether or where to partition the city or county. Is a departure from the north-up convention worth the resulting confusion or annoyance? It is usually not okay for North and South Main Streets to run toward the upper right and lower left corners, respectively, unless they are so positioned on the ground. Above all, type must be legible, and lines representing streets should not interfere with street names.

Linking the map's lines and labels to the alphabetical list of street names typically depends on a rectangular index grid that divides the map into square cells, identified by a letter and a number. The grid's columns might be labeled alphabetically starting with A at the left and its rows numbered downward from 1 at the top, so that cell "C–7" would be three columns in and seven rows down. If a street crosses multiple cells, the index might point out the cell encompassing the beginning or largest part of the street name. Each cell is a neighborhood of sort, although the cells' straight-line boundaries almost always slice through homogeneous neighborhoods.

Multiple excerpts clipped from *Clason's Guide Map of Denver, Colorado* (fig. 6), published in 1917 (and other years), show that the index grid need not be inscribed directly on the map.[31] At center-left an excerpt from the street map covers a square area roughly 1.1 miles on a side. It sits between excerpts from the letter strips along the map's top and bottom edges and between excerpts from the number strips along the map's left and right edges. At the upper right is an excerpt from the index, which lists streets by name, alphabetically, along with a pair of grid coordinates. For streets aligned with one of the cardinal directions—the dominant pattern in this part of Denver—the grid reference is either two identical numbers for a street running east–west or two identical letters for a street running north–south. For example, Gilpin Street, a north–south street referenced as "K-K," is clearly in the K column.

Anyone with excellent vision or a magnifying lens could use the Clason map's index and virtual grid to locate a specific block. For example, a user interested in 2825 Gilpin Street could look for the block number "2800" attached to York Street, just north of the route 28 streetcar line (which runs along 28th Avenue, labeled outside the excerpt area to the right). Because of consistent address numbering, the savvy user could easily figure out that the 2800 block of Gilpin was six blocks to the west of the 2800 block of York. Once on the block, house numbers made it easy to find 2825, which faces Fuller Park (one of many parks, large and small, overprinted in transparent green). Tiny labels also reveal several other destinations in the vicinity; for instance, Manual High School (a block and a half south and a block west) and the East Side Christian Church (a block and a half north and a block east). Although tiny type challenged the user's visual acuity, Clason obviously considered these additional features worth having—another of the many design tradeoffs confronting the map publisher.

House numbering and street addressing conventions like these gave city residents the real address that their country cousins lacked. Although the Post Office's RFD service, introduced in the 1890s, was a huge step in reducing rural isolation, an RFD number was good only for delivering mail, not for helping travelers find farms. Moreover, a farmer selling livestock, which the buyer might need to see, could be at a distinct disadvantage, as J. B. Plato soon discovered.

2 DENVER

Few names are as catchy as John Byron Plato, a moniker invoking key figures in apostolic Christianity, romantic poetry, and classical philosophy. That "J. B.," as he often self-identified, was not a major figure in cartography and mapping is underscored by his omission from the million-word encyclopedia *Cartography in the Twentieth Century*, published in 2015 as Volume Six of the massive *History of Cartography*.[1] This oversight is embarrassing because not only was I the volume's editor but also I had recommended decorating the first page of the *History* project's Fall 2012 newsletter with diagrams that Plato had used to explain a clever georeferencing invention he dubbed the Clock System.[2] Although his invention might be dismissed as a mere curiosity, Plato's life story and his development of the Clock System offer insight into the map's diverse roles in everyday navigation and community identity. More than a fascinating saga of one innovator's talent and tenacity, his story offers insight to the isolation of the American farmer in the early twentieth century.

Clock System cartography was Plato's response to a wayfinding problem that had little effect on his life until he became a livestock farmer in his late thirties. Born in 1876, he had first-hand experience with several transportation revolutions, namely, the fuller expansion of the nation's steam railway network, the rise of the electric interurban railway, and the development of the motor vehicle from a cumbersome curiosity to an everyday necessity for most rural residents. A farmer who acquired a tractor soon coveted a truck, which expanded their travel horizon beyond the local community center, its store, and its church. A truck not only expedited movement of farm products to a local market, canning factory, milk co-op, or train station, it also facilitated the purchase of eggs or livestock from other farmers.

My choice of Clason's 1917 Denver map as an exemplary indexed street map (fig. 6) was deliberate. J. B. Plato spent much of his early life in Denver, and 2825 Gilpin Street was his maternal grandmother's home, where he lived from at least his pre-teens through his early twenties, when he built his own home several blocks away.[3]

Plato inherited his iconic name from his father, John Byron Plato (1842–1881), and his middle name can be traced to his paternal grandfather, William Byron Plato (1810–1873). Both antecedents were lawyers: his father had earned a law degree from the University of Michigan in 1864 before establishing a successful lumber business in Denver,[4] and his grandfather had been a judge in Illinois.[5] Despite identical names, Plato and his father never self-identified as junior and senior. Around 1875 John the father had retreated to the family home in Geneva, Illinois, where the 1880 Census identified him as a "retired lawyer [with] consumption" (tuberculosis). Geneva is about thirty miles west of Chicago, where John the son was born on December 17, 1876. His father died there four and a half years later. Northeastern Illinois was the family stomping ground insofar as John senior was born in Aurora, Illinois, less than ten miles from Geneva, and his mother, Helen Francis Larrabee (1842–1931), was born in Chicago. The couple married on December 27, 1865, in Kane County, which includes Geneva. Reflecting the nineteenth century's high infant mortality, Helen Plato's two other children apparently died young, and John was raised as an only child.[6]

As lawyers, Plato's father and grandfather no doubt knew how to avoid a lengthy and costly probate by titling assets in heirs' names.[7] His grandfather left his house and its contents to his wife, and split the residuum equally among his three natural children and one adopted son, each of whom received less than $1,000.[8] His father had no will, but his mother, Helen, was appointed administratrix and received $2,919 from a life insurance policy, which was reduced to $1,331 after expenses and claims.[9] Plato himself had a simple will, which left all property and investments to his wife, whose net inheritance was valued at $61,485.[10] Whatever family money had accrued to Plato was conserved by his modest lifestyle and supplemented as needed by borrowing to buy farmland and selling shares in his businesses.

Plato's faint biographical footprint is legible largely but not exclusively in census-takers' manuscripts at the National Archives and in city directories.[11] Around 1891 he and his mother moved to Denver, where Helen had family.[12] "John B. Plato" first appeared in a Denver city directory in 1898 as a "student" residing at the same address as his mother, his maternal grandmother, and an uncle.[13] Although he probably attended elementary school in Denver, the earliest archival record of his education placed him 1,500 miles away in Ithaca, New York, in January 1896, as a nineteen-year-old student in Cornell University's "Short Winter Course" for farm youth.[14] More than two decades later Plato relocated to Ithaca, where he set up a business making wayfinding maps.

Cornell condensed the basics of farming into an eleven-week program "intended to meet the needs of those who have only the time and means to spend one or at the most two terms at the University."[15] As described in the college catalog, the program included "two hours per day of educational work in barns, dairy houses, [the] forcing house and laboratories," as well as classroom instruction in agriculture and chemistry, and in electives chosen from botany, horticulture, dairy and animal husbandry, poultry keeping, and veterinary science. Topics ranged from "farm accounts and business customs" to buildings and fences, the use of tools, and the operation and maintenance of farm machinery. Students had to be at least sixteen years old and "of good moral character." There was no tuition, and the weekly cost of room and board was estimated between $4 and $10—hardly a huge expense for Plato once he had his train ticket.

Among the eighty-three students in the Winter 1896 course, only nine were from out of state. "Plato, John Byron," from "Denver, Col." had traveled the farthest; the other eight out-of-staters were from Connecticut, New Jersey, and Pennsylvania.[16] The college yearbook listed him as an active member of the Agricultural Association, a student group.[17] The student newspaper reported him as one of six students who presented papers at the Association's March 3, 1896, meeting.[18] Plato titled his presentation, "A Voice from the West." That voice was heard again, two decades later.

Plato resumed his education in Denver at Manual Technical High School, three blocks from his grandmother's home. He attended Manual High for the two years prior to joining the First Colorado Volunteers in late spring 1898 and fighting in the Philippines during the Spanish-American War (April 1898 to August 1899).[19] He missed his last year at Manual, which until 1904 offered only a three-year program.[20] Though he was mustered out too late to see the Class of 1899 graduate, Plato remained socially involved with his former classmates, and even attended their third-year reunion in June 1902.[21]

Manual High School was a leader in the vocational education movement, and slaked Plato's interest in learning about tools, agricultural or otherwise. His two years at Manual helped prepare him for employment as a draftsman in 1900 and as owner–manager of a cabinet woods and veneer business from 1901 to 1906. Wood veneers became important in furniture manufacturing in the nineteenth century, when the Industrial Revolution expedited the cutting of logs into thin sheets of decorative hardwoods.[22] Production had increased markedly by 1906, when lumber mills in thirty states (but not Colorado) were producing hardwood veneers for use in construction and cabinetry.[23] Plato's business was apparently a retail endeavor that supplied local craftsmen from stock he kept in the backyard stable at 2825 Gilpin Street, where he lived with his mother and grandmother, four other relatives, and a family nurse. Either he sold veneers to local cabinetmakers or carpenters who glued them to a thicker wood base or he glued them himself: the process was straightforward and did not require elaborate machinery.[24] The veneers business remained at the Gilpin address even after Plato moved five blocks away to 2416 Williams Street, a one-and-a-half-story brick residence built in 1902, where he lived with his mother for the next seven years.[25]

Gilpin Street, Williams Street, and Manual High are three of several Denver locations relevant to Plato's life as a young adult. Because the sites are interconnected, I put all of them plus a few frame-of-reference features on a map (fig. 7) no larger than necessary to span a book page without undue crowding. A map showing all blocks and street names extant in 1900 provided a suitable base map for plotting locations identified by address, using city directories and a trio of more or less contemporary large-scale fire insurance and real estate atlases.[26] Reducing the opacity of the map's line symbols and street names not only allowed fully legible, high-contrast symbols and labels but also provided historical ambiance—look closely, and you can see the network of streetcar lines that converge downtown between the Union Depot, the State Capitol, and the main Post Office.

Detailed maps prepared for the insurance and real estate industries afford a graphic comparison (fig. 8) of the Gilpin and Williams properties and their neighbors.[27] Houses on both parcels were built of brick, at that time the dominant material in Denver for residential construction. In accord with the needs of an extended family, the Gilpin residence was larger than most homes on its block. Built around 1890, it consisted of a two-story front

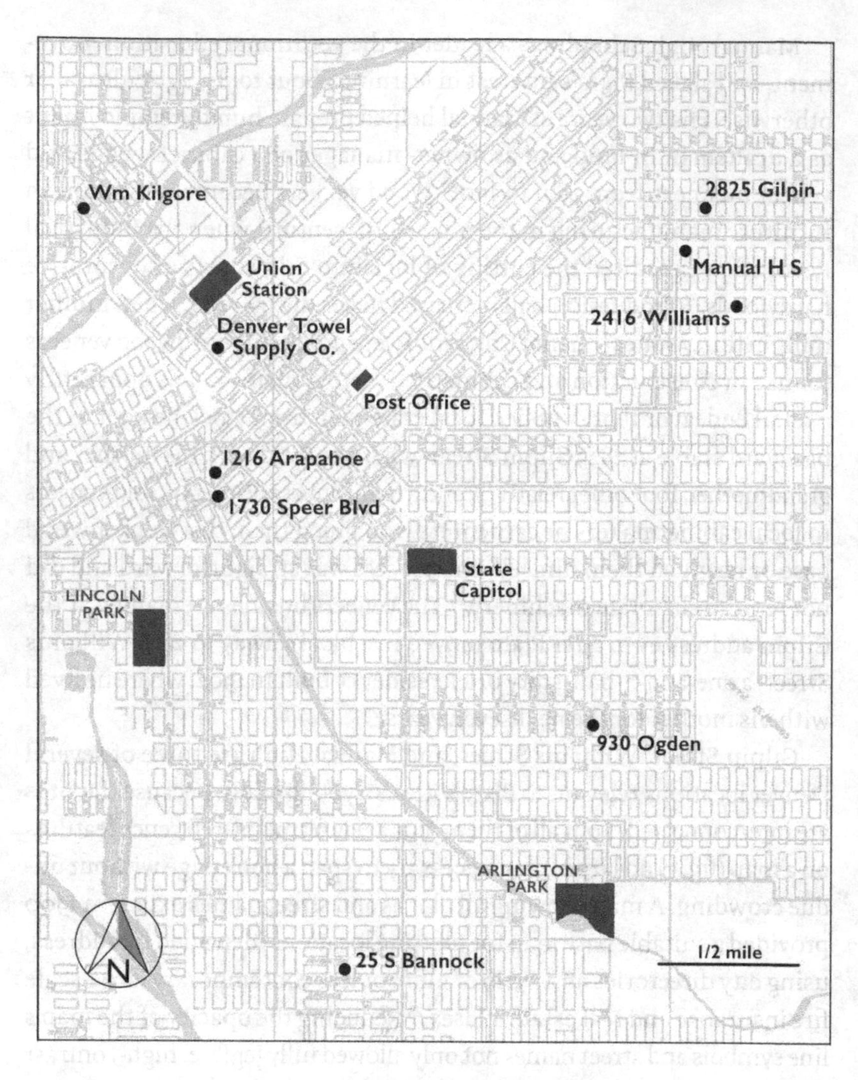

FIGURE 7. John Byron Plato's residential and work locations in Denver, 1896–1916. Data compiled from city directories, and real estate and fire-insurance atlases, and plotted on a portion of Denver Chamber of Commerce and Board of Trade, *A Map of Denver* (1900).

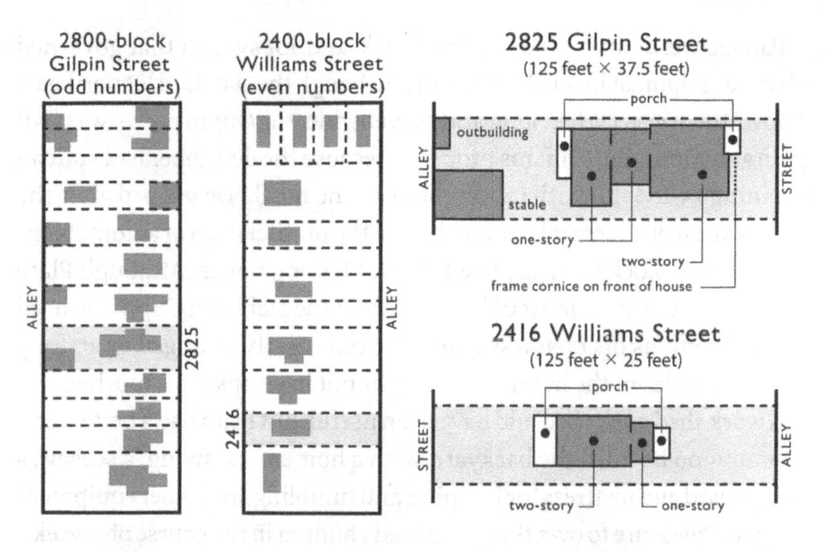

FIGURE 8. Cartographic comparison of the Gilpin and Williams Street properties describes their blockfaces (half blocks), land parcels, and structures. Maps on the left are based on a real-estate atlas, and diagrams on the right are based on a fire-insurance atlas. The 2825 Gilpin Street delineations were compiled from Sanborn's 1903–4 Denver atlas, vol. 2, sheet 220. Because the 1903–4 atlas did not include 2416 Williams Street, the latter property's delineations were based on Sanborn's 1929–30 Denver atlas, vol. 3, sheet 311.

portion, toward the street, and two smaller one-story sections. Across the street, Fuller Park provided a full block of open space, and the brick stable in the backyard might once have accommodated a horse and buggy—a less useful adjunct after Denver's expanding streetcar network connected the neighborhood with downtown in 1889.[28] The bar scale of the fire insurance map used in compiling figure 8 suggests the stable was approximately 18 × 25 feet (450 ft^2), larger than the other two stables on the block and fully adequate for a modest stock of veneers.

The small structure in the other corner of the lot, along the alley, was probably built as an outhouse. In 1890 that part of Gilpin Street was on the rural edge of a rapidly growing city, and the sewer lines had not yet caught up with the water mains.[29] Homeowners who eagerly tapped the city water system did not always add flush toilets as soon as the sewer line reached their street.

Parcels were constrained by the block-and-lot system that governed urban development in much of Denver. Although the standard lot northeast of downtown was 125 feet long and 25 feet wide, the Gilpin parcel was half again as wide as the Williams property because the developer had split an adjoining 25-foot lot with its neighbor to the north. Developed after the densification of streetcar lines and before the proliferation of automobiles, the Williams block had little need for stables or garages. Although Plato probably stored all or most of his veneers at the stable on Gilpin, he used 2416 Williams as his business address. Some nicely arranged or artfully hidden samples in the living room might not have upset his mother.

A work shed out back could have been useful, but Plato had a better idea. Around 1906 he filled the backyard with a homemade swing, a seesaw, a springboard and mattress for jumping and tumbling, and other equipment that gave "pleasure to over three hundred children in the course of a week," according to a *Denver Post* article headlined "Bachelor Makes Children's Playground of Back Yard."[30] The reporter estimated attendance from a register in which children wrote their names and ages. The playground was available after school Monday through Friday and all day Saturday, with injuries and mayhem avoided by reserving Mondays for boys aged eight to fourteen, Tuesdays for girls eight to thirteen, Wednesdays for boys four to seven, Thursdays for girls four to seven, Saturday mornings for boys eight to twelve, and Saturday afternoons for girls eight to thirteen. A gender preference that allocated Fridays to "girls who can't come on Saturday" reemerged decades later, when Plato, in his seventies and eighties, ran a campground and nature preserve outside Washington, DC, for local Girl Scouts.

Plato's backyard playground suggests not only kindheartedness but ample free time for helping his mother supervise their little guests and demonstrating the playground "stunts" mentioned in the newspaper article. The project's initiation around 1906 coincides tellingly with his shift from selling veneers to manufacturing a patented horse-hitch at a work site two miles to the southwest at 1216 Arapahoe Street. The horse-hitch was akin to a parking brake for buckboards and freight wagons: it kept the horse(s) from moving forward when a hitching post was not available. It anticipated by more than half a century the modern Amish wagon, which can be surprisingly upscale, with drum or disc brakes for stopping or parking, as well as head or taillights, speedometers, and even cupholders.[31]

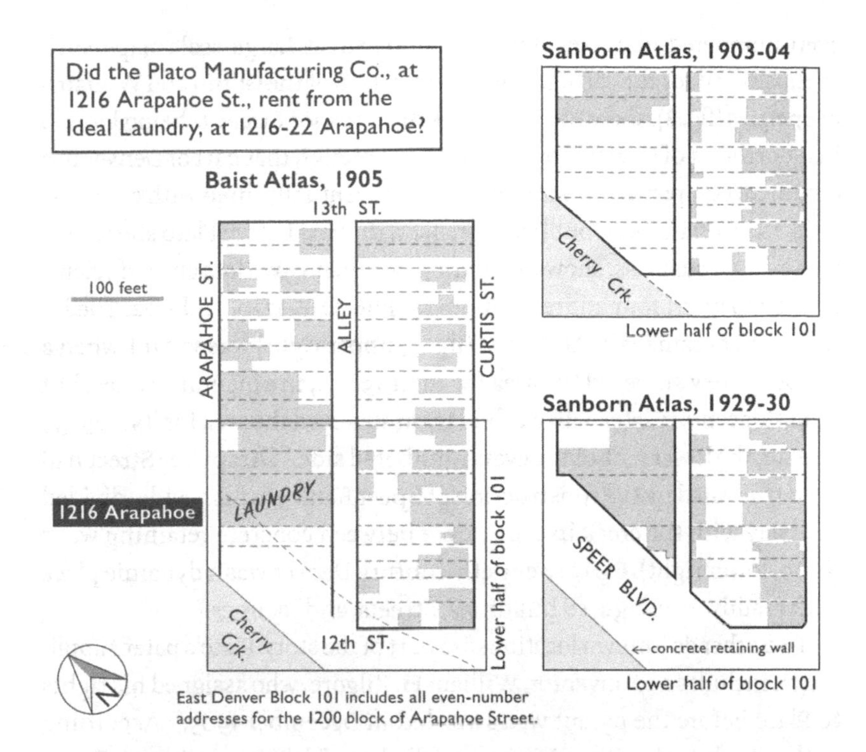

FIGURE 9. The 1216 Arapahoe Street address of the Plato Manufacturing Company refers to a laundry building in which he apparently rented space. The building did not exist in 1903 and had been partly removed by 1929. Delineations for the left panel are from *Baist's Real Estate Atlas of Surveys of Denver, Col.* (1905), plate 2. Upper right panel is from Sanborn's 1903–4 Denver atlas, vol. 1, sheet 14, and the lower right panel is from Sanborn's 1929–30 Denver atlas, vol. 1, sheet 82.

To exploit the patent through his own company, Plato rented space in a large building that housed the Ideal Laundry and the Troy Laundry, a pair of firms that shared the same president, same phone number, and same compound address: 1216–22 Arapahoe (fig. 9, left).[32] At number 1216 Plato's business would have been toward the lower, southwestern end of the street. His manufacturing business was probably not as large or substantial as his city directory listing "mgr, Plato Mfg Co" might imply.

As figure 9 illustrates, 1216 Arapahoe Street has a complicated history, confirmed by the three panels compiled from sources that bracket the year 1906. All three panels describe the lower part of a block that included even-numbered addresses along Arapahoe Street. Large-scale maps, such as those used for the cartographic comparison of the Gilpin and Williams properties (fig. 8), show the 1200 block of Arapahoe Street as being located just northeast of Cherry Creek, which ran through that part of Denver in a confined, comparatively straight channel eventually lined with concrete. *Baist's Real Estate Atlas*, published in 1905, the year before Plato started his horse-hitch business, shows a building occupying two lots and sufficiently large to merit four address numbers: 1216, 1218, 1220, and 1222. The lot that soon became 1216 Arapahoe (fig. 9, upper right) was vacant when a Sanborn crew surveyed the area for their 1903–4 fire insurance atlas, but a quarter century later, when a field team surveyed the area for its 1929–30 update, the lower part of the even-numbered side of Arapahoe Street had been truncated by the construction of Speer Boulevard as a wide, divided roadway with the creek in the middle between concrete retaining walls (fig. 9, lower right). Early twentieth-century Denver was a dynamic place with readily reconfigured buildings, streets, and lot lines.

Two other downtown locations are part of the story. Plato's patent application includes a co-inventor, William H. Kilgore, who assigned his rights to Plato before the patent was awarded in December 1905.[33] According to the 1903 city directory, Kilgore worked as a "driver" for Denver Towel Supply, two blocks southeast of the railroad tracks in front of Denver's Union Station, but lived on the other side of the tracks—figuratively and literally—in rented rooms at 1128 15th Street (fig. 7).[34] Kilgore, who had recently divorced, changed his address every year or two and might have been close to Plato's age.[35]

Three additional facts round out a set of dots begging for connection: (1) Kilgore's job involved using a horse and wagon to deliver clean towels to customers and retrieve soiled ones; (2) the wagon had to pause frequently while its driver went inside; and (3) Plato's uncle, Philip Larrabee, was the company's treasurer.[36] Kilgore and Plato probably knew each other through Uncle Phil, and Kilgore's likely frustration with a horse that would not stay put was probably a recurrent problem for which Plato had a solution.

No other explanation seems remotely plausible. Kilgore had experience with horses, Plato had the technical savvy, and Phillip Larrabee not only knew both men but might have told his nephew that the Ideal and Troy Laundries had extra space in their new building on Arapahoe Street. Although K precedes P in the alphabet, Plato's name was first on the patent for a reason.

Kilgore understood horses and wagons, but aside from helping Plato develop a prototype, he apparently had no interest in perfecting or marketing their invention. Clearly, Plato moved the invention forward by hiring Gerrit Jan Rollandet, a jack of many trades who worked as a patent attorney and professional engineer in offices across from the Post Office (fig. 7). Rollandet ran numerous related businesses at the same address and advertised in the business section of the city directory under multiple headings:

> BLUE PRINT PAPER, BLUE PRINTING, DRAUGHTSMEN, DRAWINGS—FOREIGN AND CAVEATS, ENGINEERS—MECHANICAL, MAP DESIGNERS, MAP MOUNTING, MAP PUBLISHERS, MECHANICAL DRAUGHTSMEN, MECHANICAL ENGINEERS, PATENT ATTORNEYS AND SOLICITORS, PATENT OFFICE DRAWINGS, PATENT OFFICE DRAWINGS—FOREIGN AND CAVEATS, PATENT SOLICITORS, [AND (SIMPLY)] PATENTS.[37]

Though I have no evidence, Plato might even have been a Rollandet employee in June 1900, when the census taker came calling.

Drafting played a key role in describing an invention to patent examiners in Washington, DC and to anyone wanting to license or improve upon a patent.[38] Patent applications followed a prescribed format in which one or more carefully crafted line drawings preceded a precise but often tedious verbal description of the invention's intent, operation, and key components. Numbers linked elements in the graphic description to specific words and phrases in the accompanying verbal description, known as the *specification*. A competent patent attorney had to understand both the insider's language of the specification (which I call "patentese") and the mechanical drawings that described the size, shape, placement, and interrelationships of the invention's moving and stationary parts. In the bureaucratic arena for vetting applications and awarding patents, a patent application was a coded communication between an inventor eager to convert a clever idea into a property right and a patent examiner committed to ensuring clarity and originality. Because the typical inventor was not skilled in patentese, the attorney often served as a ghost writer, while the patent examiner played the dual roles of editor and expert reviewer.

Inventors commonly assert multiple claims, and examiners typically reject at least one claim because of unclear wording, interference with another inventor's patent, or unacceptable obviousness. Through his attorney, an inventor can respond to a letter of rejection by withdrawing or amending unacceptable claims, or by contesting the examiner's conclusion, which can lead to an involved series of letters in which forced politeness masks growing frustration on both sides.[39] Though multiple rounds of rejection and amendment were common, Plato and Kilgore's patent was approved expeditiously, less than six months after filing, with only a single rejection/amendment exchange and a reduction in the number of claims from twelve to eleven.[40]

As approved and published by the Patent Office, their patent consisted of two pages of drawings followed by a four-page specification. After naming the inventors and their legal residence, the specification provided an overview in which the first of its eleven numbered figures, at the top of the first page (fig. 10), underscored the key words of the patent's title, "Hitching Attachment for Vehicles." In drafting parlance, this initial figure, which projected the side of a small freight wagon onto a vertical plane, is called a *side-view elevation*. At the front of the wagon is a buckboard, intended to shield the driver from the horse's rear hooves and whatever debris might be kicked up, and behind the driver's seat is a bed with a load that could be coal, laundry, or any of dozens of products delivered to homes or businesses. The back wheel in the foreground has been removed to reveal details below the rear axle, and dashed lines describe parts of the other rear wheel, on the far side of the bed.

The details demonstrate Plato's ingenuity. In accord with Patent Office requirements, numbers link parts of the illustration to terms in the specification. Attached to the side of the wagon, "the operating lever 16, located near the seat within reach of the driver, is fulcrumed at 21 to the side of the wagon and held against lateral movement by a segmental guide 22."[41] The drawing's "Fig. 11," barely visible at the bottom center of my figure 10, shows "notches 23" with which the lever "may be held in any desired position." The lever is connected to the apparatus beneath the "rear axle 7" by "rod 17," and the "hitching strap 37" runs forward to connect with the harness and bit. In the middle of the page, the drawing's "Fig. 2" describes "winding-drum 5," which revolves when the lever is pulled forward to engage "gear wheel 24" and "cam gear 25" with the "large gear-wheel 29" on the wagon wheel. Thus engaged, the winding drum tightens the strap and reins in the horse. "An additional but not

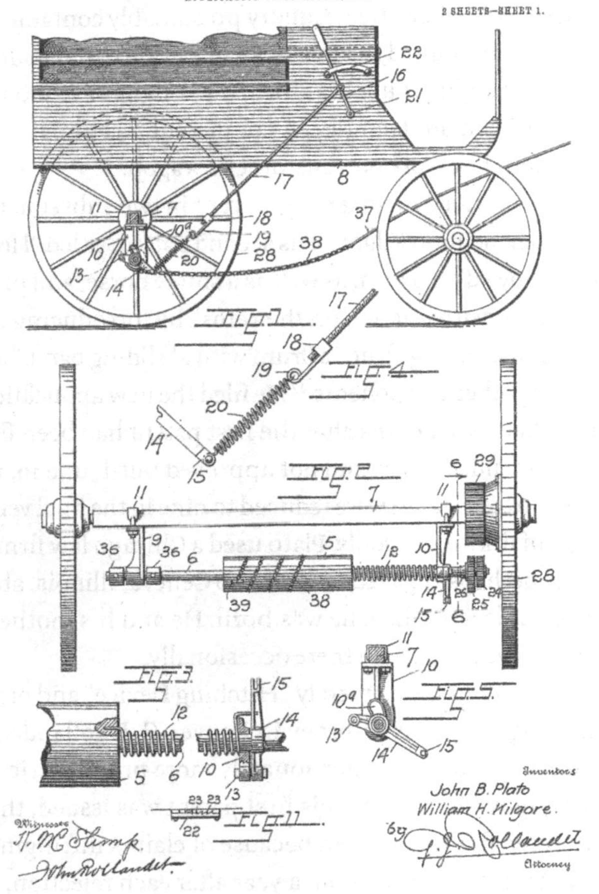

FIGURE 10. First page of the horse-hitch patent, filed April 29, 1905, and issued December 6, 1905. J. B. Plato and W. H. Kilgore, "Hitching Attachment for Vehicles," US Patent 807,047, filed April 29, 1905.

less important" part of the invention loosens the tension on the strap when the horse backs up. Clever.

I found no evidence that Plato licensed the invention to a wagon manufacturer, franchised his components and their assembly to another dealer, or advertised in a trade journal or a local newspaper. Word of mouth and a listing in the city directory's business section, along with a phone number, were apparently sufficient to keep the business afloat. His shop in space rented from the Arapahoe Street laundry presumably contained whatever machine tools were required to enhance an invention that remained a series of prototypes throughout its short life. Plato might have modified parts bought from other manufacturers. A competent machinist, he needed a space large enough to work on a customer's wagon.

That Plato continued to tinker is apparent in two subsequent patents, neither of which mention Kilgore. His second patent, titled "Horse Hitching Device," allowed for a team as well as a single horse, and promised an "automatic and positive action on the reins" by introducing a "toothed rim," a "rising-and-falling shaft," a drum with a "sliding band," and a "coil-spring," among other components.[42] He filed the new application on June 21, 1905, less than two months after the first patent had been filed. It was rejected and amended twice, and not approved until June 19, 1906, after the original fourteen claims were reduced to nine in the final version, with a single page of drawings. Oddly, Plato used a Chicago law firm, Banning and Banning, perhaps engaged on a visit to Geneva, Illinois, about thirty miles west of Chicago, where he was born. He and his mother still had relatives in Geneva and visited there occasionally.

Plato's third patent, titled merely "Hitching Device" and orchestrated through his original patent attorney in Denver, G. J. Rollandet, reflected further tinkering and was proportionately more problematic.[43] Filed on December 25, 1905, shortly after his first patent was issued, the application was rejected five times, often because of claims infringing another inventor's patent. A cadence set in: a year after each rejection, Rollandet submitted an amended application, which was rejected three to six weeks later, and so on.[44] The cycle was broken when a July 26, 1910, examiner's letter merely correcting several phrases in the specification and inconsistencies in two of the figures was followed promptly, on August 6, 1910, by an amended application, and on August 18, 1910, by the examiner's letter of allowance. Plato's twenty-six original claims had been whittled down to a mere nine, but because the final fee was not paid promptly, the patent was not issued until March 14, 1911. By this time, he had probably lost interest in his invention.

Figure 11 supplements the cartographic summary in figure 7 by imposing a temporal order on the various places, events, and experiences of

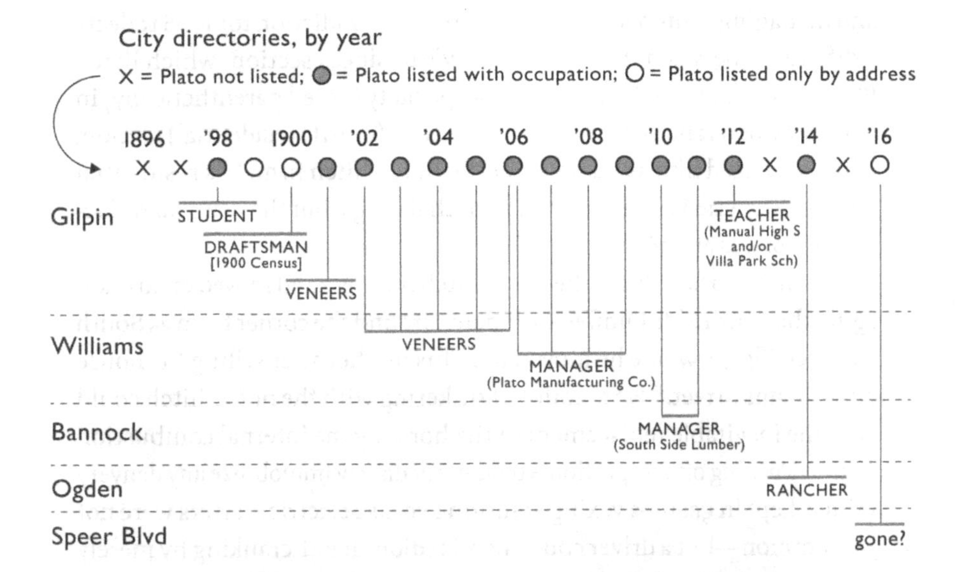

FIGURE 11. Timeline for Plato's residential and work locations in Denver, 1896–1916. Columns represent city directories, by year, and rows show place of residence (left) and occupation. Data compiled from *Ballenger & Richards Annual Denver City Directory* (various years), *Baist's Real Estate Atlas of Surveys of Denver, Col.* (1905), and *Sanborn's Atlas of Denver, Colorado* (1903–04 and 1929–30).

Plato's life in Denver between 1896, when he returned from Cornell, and 1916, when he was firmly settled on a farm in Semper, nine miles north of Denver, as discussed in the next chapter. I used Denver city directories to compile a timeline that relates their year of publication to Plato's reported type of employment and place of residence.[45] The graph's rows indicate where he was living, and its columns either identify his occupation or mark the years for which he was either not listed or no occupation was reported. That the directories for 1899 and 1900 omitted an occupation probably reflects Plato's service overseas, in the Philippines during the Spanish-American War; his mother and other relatives living at 2825 Gilpin Street would have confirmed their home as his permanent residence. Even though the 1900 directory did not report an occupation, the manuscript schedule for the 1900 Census identified him as a draftsman, at least on June 4, when the census taker visited. In columns to the right, the graph reflects his business experience selling veneers, manufacturing the horse-hitch, and managing a lumberyard. An apparent contradiction for 1906 reflects a difference between the city directory's business section, which listed Plato under "Lumber Dealers" with a specialty (noted parenthetically) in "cabinet woods and veneers," and its general (mostly residential) section, which reported him as manager of the horse-hitch firm. His residential address was 2416 Williams Street in both listings, but the Plato Manufacturing Co. was at 1216 Arapahoe.[46]

By 1910 Plato had closed the horse-hitch business and moved on to managing the South Side Lumber Company, around the corner from 25 South Bannock (fig. 7), where he moved with his mother after selling his house on Williams Street.[47] No amount of tinkering with the horse-hitch could alter the inevitable replacement of the horse by the internal combustion engine. Turning off the ignition would immediately immobilize any delivery vehicle kept in gear—a vexing solution because electric starters were not yet common—but a driver could avoid tedious hand-cranking by merely depressing the clutch, putting the transmission in neutral, and setting the parking brake.[48] Rising sales of motor trucks and buses, up abruptly from 750 in 1905, to 6,000 in 1910, and 74,000 in 1915, underscored the eroding market for Plato's invention.[49]

Although his father's prominence in the 1870s as a Denver lumber retailer and his own experience in marketing veneers might account for the move, running a lumberyard was apparently less fulfilling than farming, the basics of which he had learned a decade and a half earlier during three memorable months at Cornell. He must have been exploring a new career in agriculture as early as 1909 because in January 1910 he paid $800 for two adjoining five-acre rectangular lots in Semper, a semiarid and largely undeveloped community seven miles north of the Denver County line.[50] Semper was named for a couple who had settled there around 1880, encouraged the digging of an irrigation canal, and fostered the development of an agricultural community by donating land for a school.[51] Land speculation encouraged further settlement.

Figure 11 has another ambiguity: the 1912 city directory listed Plato as a teacher at Villa Park, a public high school off the map about two miles southwest of downtown, whereas in early September 1911 the city's two daily newspapers, which published lists of teachers and their assignments (provided by Denver school officials), placed him at Manual High School, which he had attended before joining the First Colorado Volunteers.[52] This is not an inherent contradiction insofar as the directory could reflect a half-year appointment at Villa Park in spring 1912 following a half or full year on staff at Manual. Teaching would seem a convenient transitional job when Plato was shifting from manufacturing to farming, and he no doubt had considerable practical experience relevant to teaching manual arts.

The time-series graph also reflects Plato's erratic residence in Denver after 1912. Although a "rancher" living within the city in 1914 might seem farfetched, maintaining a downtown pied-à-terre would have been useful in mid-decade, when he was making and publishing maps as well as running a small dairy and livestock farm outside Denver. High-speed electric railcars that ran every hour between Denver and Boulder on the Denver & Interurban Railroad not only stopped at the Semper depot, less than a half-mile from Plato's farm, but also originated at a trolley loop just a few blocks from his Speer Boulevard residential address in the 1916 city directory.[53] Running time was twenty-nine minutes, and the one-way fare was twenty-five cents.[54] As farmers close to an interurban stop might have appreciated, rural isolation varied with where in the country one lived.

3 SEMPER

Plato's life at Semper, Colorado, is a complex story best told with maps at two different scales. One map describes his dynamic land holdings at Semper Garden Tracts, the rural subdivision where he bought, sold, paid taxes, and defaulted on multiple five-acre lots while buying, selling, and breeding Guernsey cows. It was here in Semper, at Featherton Farm (as he dubbed it), that he recognized the lack of a "real address" as a problem and conceived the Clock System map and directory as a solution. Semper was also the base from which he filed for patent protection, tried to sell the idea to the Post Office Department, and orchestrated two prototype Clock System maps and directories. The second map, with a broader geographic scope, relates Plato's train stop in Semper to his RFD post office in Broomfield, his attorney's office in Denver, and his two cartographic prototypes, for Longmont and Fort Collins.

My first map (fig. 12) identifies the thirteen numbered lots that comprise Semper Garden Tracts. Except where truncated on the north by the right-of-way of the Colorado and Southern (C&S) Railroad, the lots are five-acre rectangles. In making the map I had to resolve discrepancies among multiple sources that differ in content, scale, time, and geographic reliability—a bloody-minded approach that passes for cartographic license. My source for the lot lines and numbers is a subdivision map drawn around 1896 by the Colorado Bond and Realty Company and provided by the Jefferson County archivist.[1] Although the developer's drawing is a reliable source for lot numbers, I had to correct the sloppy, inconsistent spacing of its horizontal lot boundaries. Field surveyors and lot buyers would not have tolerated this imprecision.

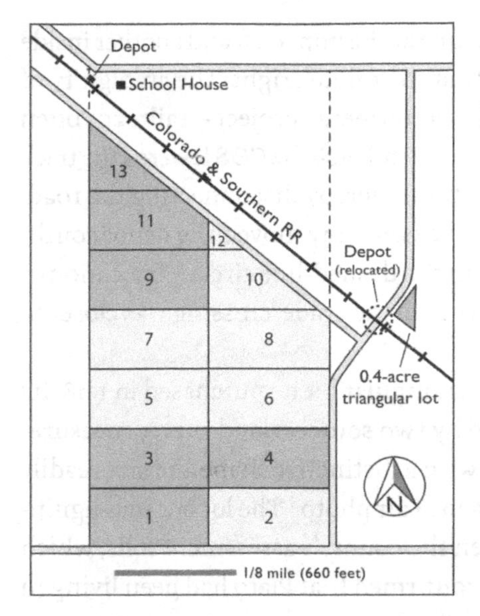

Depot
School House
Colorado & Southern RR
13
11
12
9
10
7
8
5
6
3
4
1
2
Depot
(relocated)
0.4-acre
triangular lot
N
1/8 mile (660 feet)
Semper Garden Tracts
Numbered lots of 5 acres or less
Darker shading indicates lots that Plato once owned, however briefly. After 1908 an electrified track immediately northeast of the C&S line served the Denver & Interurban RR.

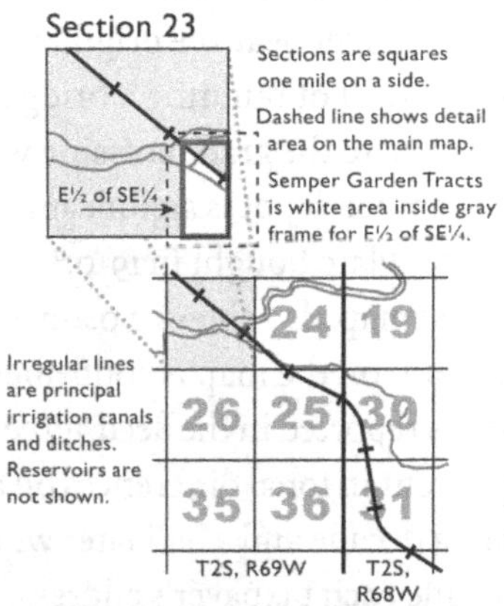

Section 23
Sections are squares one mile on a side.
Dashed line shows detail area on the main map.
Semper Garden Tracts is white area inside gray frame for E½ of SE¼.
E½ of SE¼
Irregular lines are principal irrigation canals and ditches. Reservoirs are not shown.
24
19
26
25
30
35
36
31
T2S, R69W
T2S, R68W

FIGURE 12. Semper Garden Tracts, in Jefferson County, Colorado. Numbers identify individual lots, each five acres in size, except where the subdivision was truncated by the railroad. At various times Plato owned lots 1, 3, 4, 9, 11, and 13. In 1918 he purchased the 0.4-acre parcel east of the subdivision and immediately north of the railway. Compiled from Colorado Bond and Company, *Semper Garden Tracts* [map], circa 1910, and other sources provided by the Jefferson County archivist.

The locator inset at the lower right relates the subdivision's thirteen lots to section lines of the Public Land Survey System (PLSS), which provides a convenient legal description of the square-mile section before its subdivision into smaller parcels.[2] It also relates the boundaries of section 23 and five other sections in the same township to the territory covered in greater detail on the main map.

A prominent feature on the main map is the C&S Railroad, which runs from upper left to lower right (fig. 12). From 1908 to 1926 the C&S hosted a wholly owned subsidiary, the Denver & Interurban Railroad (D&I), on an adjacent track. The C&S ran both freight and passenger trains, always steam-powered, whereas the D&I, which carried only passengers, connected Denver and Boulder with electric trains (typically a single self-propelled car, perhaps with a trailer), which ran both ways every hour at speeds approaching sixty miles per hour.[3]

Note the two depots in figure 12: one at the upper left and another inside a dashed-line circle of uncertainty at the middle right. The ambiguity of two Semper depots reflects a C&S "betterment" project—railroads often tweaked their track to fix mistakes.[4] Around 1908 the C&S lowered the track about 20 feet near the original station, thereby disconnecting the road, but instead of building a bridge, the company moved the depot southeastward to the grade crossing with the dashed-line circle.[5] Relocation of the Semper depot is significant because the grade-crossing was closer to the lots Plato bought in 1910.[6]

The map also shows a 0.4-acre triangular lot he purchased in 1918. Its position on the map is confirmed by two sources: land survey measurements reported in the deed agree with a distinctive shape and size readily apparent in tonal differences on a 1956 air photo.[7] The lot became significant a decade and a half later when the county's assessment rolls, which include each taxpayer's address, confirmed that Plato had been living in Ohio after leaving Ithaca.

Only the inset map shows the area's principal irrigation canals, based largely on the topographic map. Although irrigation ditches and canals would clutter the main map with intricate details largely peripheral to Plato's story, his land included water rights, essential for a small farm in a semiarid area otherwise suitable only for ranching or dry farming, which required much larger parcels.[8]

Was an irrigated five-acre lot sufficiently large for a farm? The Colorado Bond and Realty Company, which platted the trapezoidal subdivision, apparently thought so.[9] With downtown Denver nine miles away, Semper was convenient for farmers eager to sell their milk, eggs, fruit, and poultry to consumers in a growing city. With the arrival of the D&I in 1908, Semper Garden Tracts was particularly well suited to a farmer with relatives, a teaching job, and business connections in the city. Even so, Plato was not content with only two five-acre lots, numbers 1 and 3 (fig. 12). In early 1911 he paid $1,650 for lot 4, adjacent to lot 3, but sold it back two years later for the same amount.[10] Still eager for additional land, he acquired lots 9, 11, and 13 in March 1916, for a mortgage of $2,500.[11] Contiguous to each other but not to his original ten acres, these lots were only a quarter mile away along a right of way between intervening lots.[12]

Tax records offer insight to how Plato used his lots. Every year the county assessor posted separate valuations for land parcels, improvements thereon, and the owner's personal property. (In central Colorado, water rights were treated as a separate land assessment based on the owner's fractional share of a reservoir with a particular acreage.) As a category, *improvements* included structures like barns, silos, and farmhouses, whereas *personal property* was anything, tangible or intangible, that was not tied down and could be bought and sold.[13] On a farm, personal property included livestock and agricultural machinery as well as furniture, wagons, cars, and trucks.

Plato's yearly assessments suggest he never built anything more substantial than a small barn on any of his lots. His assessment for improvements to his original property (lots 1 and 3) jumped from $0 in 1913 to $150 in 1914, which probably reflects the 80 × 16-foot structure he built for his eleven dairy cattle and described in a short article in the May 1917 issue of *System on the Farm*.[14] His outlay for materials was a mere $60, a tenth the price of a conventional barn, he claimed. One of its long sides was covered only by canvas, which was easy to clean and convenient for unloading straw. The canvas cost only $3.50—Plato bragged about spotting a bargain in town—and tearing was unlikely because he had cut off the horns of all but two of his cattle. "But even those with horns do not damage the canvas [because] they quickly learn to lower their heads and to walk right under."[15] The assessed improvement for the other, more northern part of his farm (tracts 9, 11, and 13) stabilized at $100 the year after its purchase—hardly enough to cover more than fencing and a storage building. Assessment rolls for 1913 and 1914 suggest that a typical small farmhouse would have raised his assessment by at least $300. Although Plato could have bedded down with the cattle in a pinch—he claimed the shed "could handle at least twenty cows . . . if I had them"—I assume he rented a room somewhere in the neighborhood or built a small, one-room shelter to supplement a rented room or hospitable relatives down in Denver. Despite ready access to the city, Plato was sufficiently engaged in his local community to win a seat on the Semper school board in 1916.[16]

Additional evidence confirms that Plato, however small his acreage, was a serious livestock farmer. His assessment for personal property, which jumped from $80 for 1913 to $130 for 1914, and escalated to $530, $570, $650, and $1,130 for subsequent years, suggests a progressive accumulation of capital, though the latter increase, between 1917 and 1918, might have included an automobile as well as livestock.

Although available archival data, unfortunately, do not include the assessor's notes or motor vehicle records, advertisements in local newspapers and breeder's periodicals indicate an ongoing effort to expand and improve his small herd. In May 1914, for instance, he advertised a "FRESH cow with fine veal calf; both for $65" in the *Denver Post*, and in August he "WANTED—Several extra good Guernsey or Jersey cows."[17] In a short piece Plato published in 1914 in the *Rural New-Yorker*, a weekly newspaper for farmers, he expressed a preference for the Guernsey, which is "a more profitable animal than the average Jersey [because] a certain sum invested in a Guernsey will buy a higher class animal than will be the same amount spent for a Jersey."[18] Like pedigreed dogs and cats, which have distinctive names that include the original breeder's kennel or cattery, purebred cattle have names like Flossie of the Bar Forks 39548 and Lily III of the Brickfield 28375, which Plato bought in 1918, and Pioneer's Dairymaid of Featherton Farm and Pioneer's Goodie of Featherton Farm, which he sold in 1920.[19]

Plato depended on classified ads to attract a buyer's interest, on the mails to arrange a visit, and on the visit to make the sale. He also depended on journalists to spread the word of his invention, as in an interview published in the April 1917 issue of *Illustrated World*, a monthly technology magazine.[20] The article recalled a seminal moment in 1914, when his classified ad attracted the attention of some "eastern buyers"—*eastern* within the greater Denver area, that is—who wrote to arrange a visit. His address at the time was Featherton Farm, Route 1, Box 112, Broomfield, Colorado, which worked fine for an exchange of letters but proved useless when the would-be buyers could not get directions.[21] They asked at the local post office, which did not have a map. Advised to wait until the next day and follow the letter carrier on his route, they returned home and mailed a polite apology, with regrets that "It took time to find your place." An equally frustrated Plato was annoyed that his postal address had "killed a mighty profitable bargain." Having lived for decades in a city, he was amazed that a rural resident "could submit to having his place lost under an R. F. D. alias that only he and the mail carrier could decipher."

The incident started him thinking: "'Takes time to find your house'... kept running through my mind. And, of course, every time that phrase came to mind there appeared a mental picture of a watch or clock—and that was my solution. I realized it one day when my own watch had stopped. As I stood scowling into its face it smiled back my answer—the solution

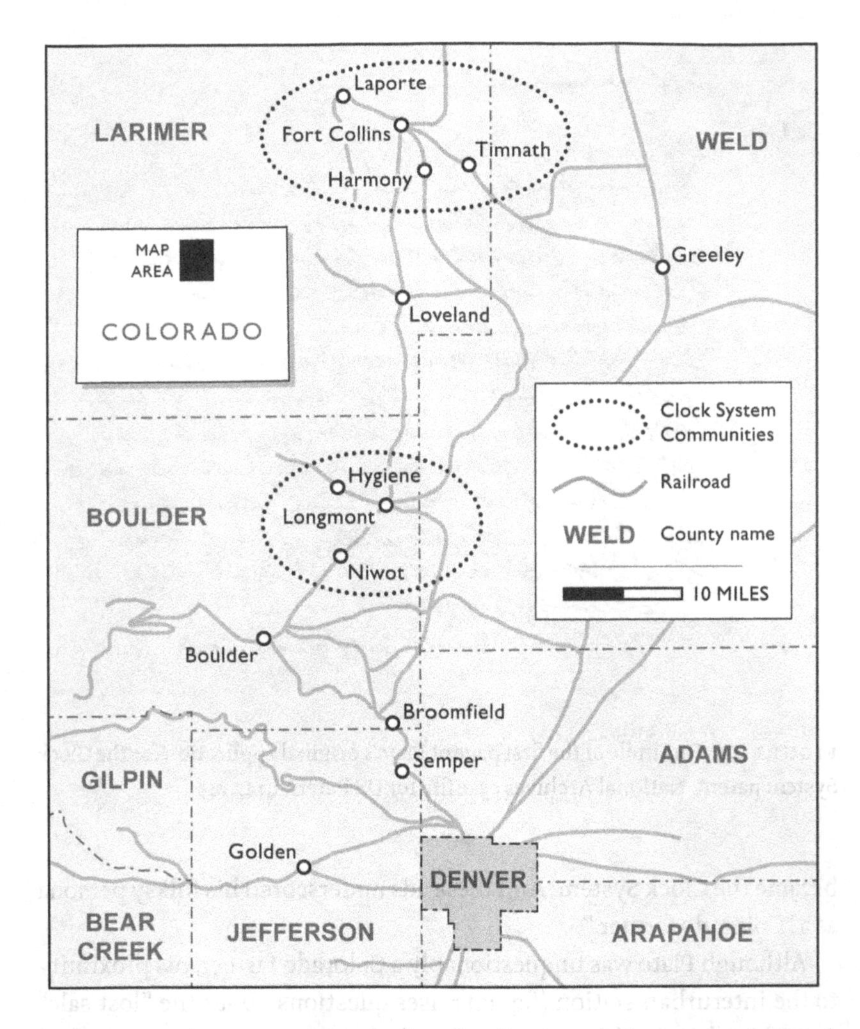

FIGURE 13. Places around Denver that Plato lived, worked, or mapped. Dotted lines encircle cities and towns named in the title of a Clock System map. Compiled by the author.

of my problem. It was a key that everyone carried, it was familiar to the smallest child."[22] The clock face divided the area around a local community center into twelve sectors, which concentric circles could subdivide into "blocks," and within each block a letter or number could uniquely identify every farmstead. A habitual storyteller, Plato repeated this tale with minor variations in numerous interviews as well as in ads for what

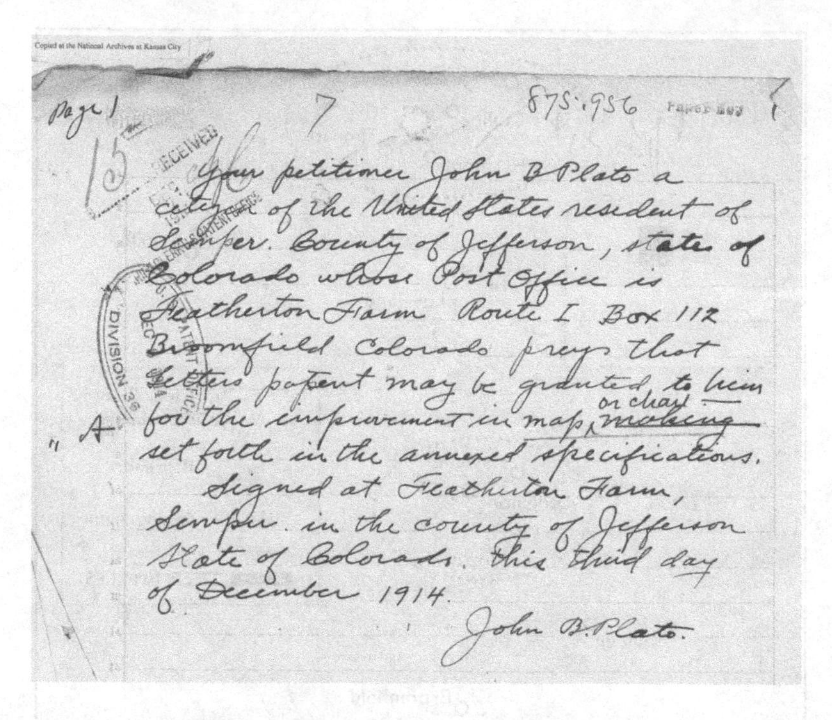

Copied at the National Archives at Kansas City

Page 1 7 875.956

Your petitioner John B Plato a citizen of the United States resident of Semper. County of Jefferson, state of Colorado whose Post Office is Featherton Farm Route I Box 112 Broomfield Colorado prays that Letters patent may be granted to him for the improvement in map or chart making set forth in the annexed specifications.

Signed at Featherton Farm, Semper. in the county of Jefferson State of Colorado. this third day of December 1914.

John B. Plato.

FIGURE 14. Facsimile of the first page of Plato's original application for the Clock System patent. National Archives case file for US Patent 1,147,749.

became the Clock System. And these ads underscored his folksy persona as a "Colorado farmer."

Although Plato was unquestionably a Colorado farmer, his proximity to the interurban station (fig. 12) raises questions about the "lost sale" narrative alleged to have inspired the Clock System. Because Semper had no post office, he received his mail from an RFD letter carrier based in Broomfield, five miles up the line from Semper (fig. 13). Why, I wonder, weren't the disappointed buyers, who "took the afternoon train back to Denver," advised beforehand to arrive at Semper on the interurban and take the short walk to Plato's small barn on lots 1 and 3?[23] Perhaps he naively saw no need for precise directions. Or maybe the eager buyers confused his postmark with the train stop when they bought their tickets. I should give Plato the benefit of the doubt because in his application for the Clock System patent he explicitly self-identified as "a resident of Semper" (fig. 14).

Whatever his inspiration, Plato wasted little time in applying for a patent. Resourceful as well as thrifty, and perhaps soured by the repeated rejections and amendments that impeded his third horse-hitch patent, he filed the application without hiring an attorney.[24] And as shown in figure 14, he wrote it longhand, using language that he apparently believed would pass for patentese.

Though his penmanship was exemplary, and he wrote on legal-size paper with numbered lines, the Patent Office was not impressed. In a typed response, patent examiner L. S. Underwood (probably not related to typewriter magnate John Thomas Underwood) observed, "The specification is not prepared in accordance with the standard required by the office and is also lacking the customary preamble."[25] A revision was needed "in such clear and definite language that a full understanding may be had of the invention and its mode of operation." Moreover, Plato had "presented claims for [an] invention drawn to a 'system,' and such claims are objectionable, in that the so-called system involves merely the idea, rather than a means for carrying out the invention."

Another concern was his application's single drawing, "received in a mutilated condition." A sympathetic Underwood deemed it acceptable "for purposes of examination" and advised encouragingly that "a new drawing will not be insisted upon until the case is in condition for final action." Because the drawing "illustrated a map or chart for carrying out the system . . . it is to this chart that claims should be drawn." With constrained optimism, he conceded "patentable novelty" and suggested that "claims properly presenting [this novelty] may receive favorable consideration."

More pointedly, Underwood advised that Plato "would best serve his own interests by seeking the aid of competent, registered patent counsel"—which he did by promptly engaging his horse-hitch attorney G. J. Rollandet from Denver. Little more than a month later, Rollandet had submitted a revised, carefully typed application with a new drawing as well as the title "Map or Chart" endorsed in Underwood's letter.[26] Despite the freshened application, the examiner rejected all seventeen claims, mostly for probable infringement on an existing patent, and returned the drawing because it failed "to properly illustrate the appearance of a map with a transparent member superposed thereon."[27] We will never know how poorly the submitted drawing treated the transparent overlay—returned by the Patent Office, the rejected artwork was never part of the National Archives' case

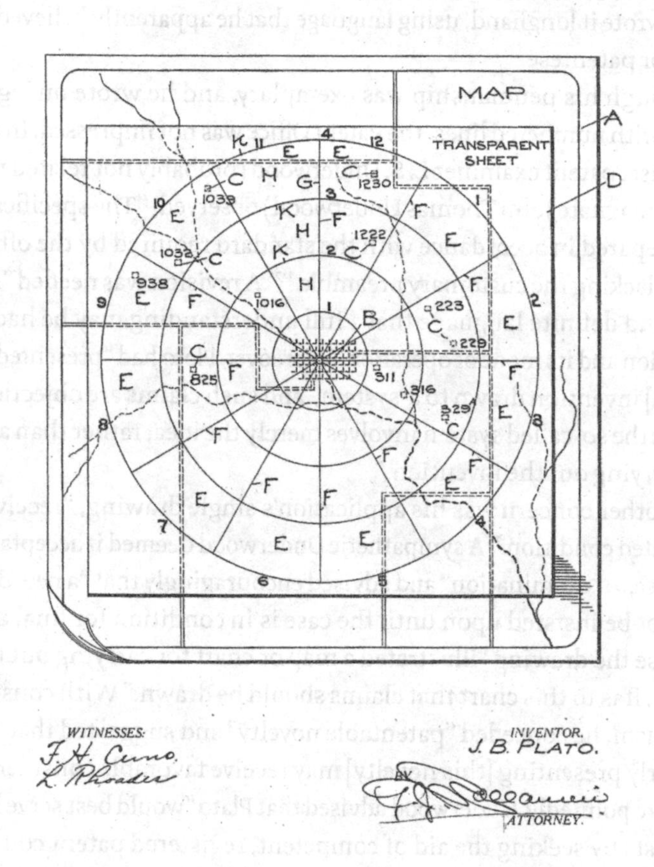

FIGURE 15. First page of the Clock System patent contains the patent's only drawing. J. B. Plato, "Map or Chart," US Patent 1,147,749, filed December 7, 1914, and issued July 27, 1915.

file—but the approved drawing uses a spiffy drop-shadow to underscore the third dimension of a movable overlay clearly labeled "Transparent Sheet" (fig. 15).[28] In addition to revising the drawing, Rollandet reworked the specification, and after another round of correspondence to tweak the final language, the eight revised claims were allowed, all fees were paid, and the patent was issued on July 27, 1915, less than eight months after Plato had mailed his handwritten application.[29]

Why the patent mentions a "transparent sheet" is not clear. Although Plato might have used a movable overlay in configuring boundaries separating a county's multiple community centers, a transparent sheet is not otherwise essential in either making or using a Clock System map. Indeed, the specification mentions it just once, as an overlay "which when laid upon the map divides the area it covers," but the same sentence immediately adds it "may be drawn or printed directly on the face of the map or chart."[30] Rollandet was probably doing what any competent patent attorney would have done to convince the Patent Office to approve an expansive set of claims that wards off whatever twists and tweaks might give a competitor or parasitic inventor a free ride on his client's ingenuity.

This strategy explains the purposeful repetition in the specification's discussion of the invention and its eight seemingly redundant claims. This deliberate verbosity is anchored by a single illustration (fig. 15) that shows all of the map's key elements: its twelve, evenly spaced radial lines that divide the area into "districts" or sectors; its evenly spaced concentric circles that divide the districts into "zones" or neighborhoods—later called "blocks"—the point symbols marking individual rural residences, each identified by a unique number (0, 1, . . . , 9) or letter within its zone; and frame-of-reference features such as roads and streams. Although several claims specifically call for "twelve lines that radiate equidistantly," claims that omit a specific number of lines implicitly include a system with only eight lines, as in the Compass System, introduced after Plato's patent had expired. Moreover, despite a general reliance on concentric circles a mile apart, three of the eight claims expansively recognize "distances expressed in miles or other units of measurement," and despite a general focus on "dwellings in rural districts," the specification notes the system "may be used equally well for the designation of mines or other geographic locations on a map or chart." And should the number of locations within a zone exceed ten, "the numbers can be further distinguished by the addition of a figure, letter, exponent or other symbol as illustrated in the drawing in the district between lines 3 and 4." Look carefully at the patent drawing and you will see "329^{A}" in district 3, zone 2. In language Plato is unlikely to have concocted by himself, Rollandet tried to cover all foreseeable variants, including dots distinguished not by single digits but by letters, which supported twenty-six locations within a Clock System neighborhood.

Conspicuously absent from the patent is any mention of a rural directory listing addresses by the householders' names or numerically, as in a city directory's crisscross section. The reason for this omission is revealed in the patent's second paragraph: "This invention relates to a system of designating dwellings in rural districts, by numbers or other distinctive symbols, principally for the purpose of facilitating the delivery of mail by what is commonly known as the rural free delivery service."[31]

That Plato eventually established a firm that produced Clock System maps was an unintended consequence of an invention originally intended to give farmers a mailing address as functionally meaningful as the city dweller's street address. His goal was a system of geographically meaningful numbers, not a map. Postmasters, not everyday citizens, would be his primary customers.

This goal explains Plato's next step: visiting Washington to sell the idea to the Postmaster General. According to his premier *Clock System Rural Index*, published in Ithaca in 1919 for the Town of Ulysses, New York, Plato had "a number of conferences [with] representatives from the Post Office Department, the Agricultural Department and the Department of the Interior"[32]—all with less success than *Illustrated World*'s assertion that "a Post Office official told him that he had exactly what the Department had been looking for for years, and that they would be glad to incorporate it in their system."[33] The official might have been encouraging but nothing happened. Although the US Postal Service's History Office had no record of Plato's having visited postal headquarters,[34] I found two newspaper stories that confirmed his business trip.

On November 21, 1915, the *Tacoma Daily Ledger*, in a report from its Washington bureau, concluded that postal officials were giving "serious consideration" to "a novel plan for numbering the rural mailboxes throughout the United States in such a way that the number on the box will tell not only the direction, but the distance, of the box from the post office out of which the rural route emanates."[35] To enhance the Clock System's purpose, Plato was now including an address sign akin to a city residence's house number: a strategy endorsed by the *Ledger* as "a great boon to motorists and others traveling through the country [because] every rural mail box would be a sign post pointing the way with unerring accuracy to the nearest town or city." Anticipating a favorable outcome, the report noted that the inventor "is here and will spend several days pointing out its advantages to post office department authorities." Plato had apparently bought his own train ticket, paid for his hotel, and arranged for someone to milk and feed his cows so he could pitch his patent in person to the postal bureaucracy.

The *Daily Ledger* must have been linked to a wire service because four days later the *Colorado Transcript* in Golden, Colorado, used identical language in reporting "the post office department is giving serious consideration to a novel plan for numbering mailboxes."[36] And in early February a follow-up article endorsed the Clock System by complaining, "Rural routes, at present, are numbered according to their seniority of establishment, and the various boxes thereon either receive no number at all or are numbered according to the whim of the rural route carrier."[37] The *Transcript* had also found a use for "the transparent celluloid key which the postmaster can lay upon the map with its center at the point of mail distribution"—a usage not specifically identified in the published patent. If the Post Office had adopted Plato's addressing scheme, he could have faced the unintended consequence of making and selling a transparent accessory not specifically protected by a claim in his patent.

Plato's penchant for talking to journalists not only confirmed his largely unproductive trip to Washington but also revealed his mapping business as a vertically integrated venture that included mailbox signs as well as directories. Indeed, the trip might not have been a complete waste, according to the vaguely worded Ulysses directory, which reported ". . . it was decided that the best and quickest way to get the 'Clock-System' into actual use was to develop a plan whereby the work could be taken up by small units, standardized, and made self-supporting."[38] I assume that Plato was one of the deciders, perhaps the only one.

For the system to be self-supporting, it needed a printed directory with its own revenue stream: advertising. Undaunted by Washington's lukewarm interest, Plato decided to compile a Clock System map and accompanying directory, and publish it with his own imprint, identified in his copyright filing (fig. 16) as the United States Rural Directory Company, Mountain States Division, based in Denver. Because the rapidly growing Denver exurbs were presumably too large or otherwise unmanageable, he chose Fort Collins, about sixty miles north (fig. 13), for his prototype.[39] In 1910 Fort Collins and Larimer County, in which it is located, had populations

FORT COLLINS, Col. Rural directory for Fort Collins, Timnath, La Porte [etc.] 1916 ed. cover-title, 8 p. fold. loose map. fol. $0.50. [306
© Nov. 25, 1916; 2 c. Dec. 7, 1916; aff. Jan. 2, 1917; A 448782; United States rural directory co., Denver.

FIGURE 16. *The Catalog of Copyright Entries* listed Plato's directory under "Pamphlets, etc." (a subcategory of "Books") rather than under "Maps," a separate category. The entry includes the copyright's registration number (A448782) and its sequence number in the *Catalog* volume for the year (306). Library of Congress, Copyright Office, *Catalog of Copyright Entries*, Part 1: Books, Group 2, n.s., vol. 14, no. 1 (Washington, DC: Government Printing Office, 1917), 13.

of 8,210 and 25,270, respectively, up markedly from counts of 3,053 and 13,168 for the 1900 Census, and sufficiently expansive to suggest a need for yearly revision.[40]

Two scanty news stories confirmed the endeavor. A September 1916 article in an Oklahoma newspaper quoted the unnamed "secretary of the American Opportunity League," who in describing Plato's system, noted that "greatly increased rural travel" called for replacing the antiquated system of mailbox numbers, and reported that "right now he is making a county map in Colorado."[41] And on Christmas Day, a Topeka, Kansas, newspaper reported that "what is known as the 'clock system' of farm numbers has been established at Fort Collins, Colo." and added that Plato's "plan to put the farmer on the map, literally . . . will be worth millions of dollars every year to the newspapers, merchants and advertisers of America."[42]

Published in late November 1916, Plato's Fort Collins map was printed on thick, stiff paper. A vertical crease down the middle divided one side into front and back covers (fig. 17), and on the opposite side the map extended across the crease like a magazine centerfold (fig. 17). The two cover panels carried ads for six firms, all pitched toward farmers and all based in Denver: one was an eager buyer of eggs and poultry (specifically chickens, turkeys, ducks, geese), and the other five offered a broad range of products: steel pipes and tanks, calf meal, crackers and cookies, paint, and overalls. In a pitch to local interests, Plato had the map printed in Fort Collins by the Courier Press, a union shop identified by the tiny logo at the bottom of the back cover.

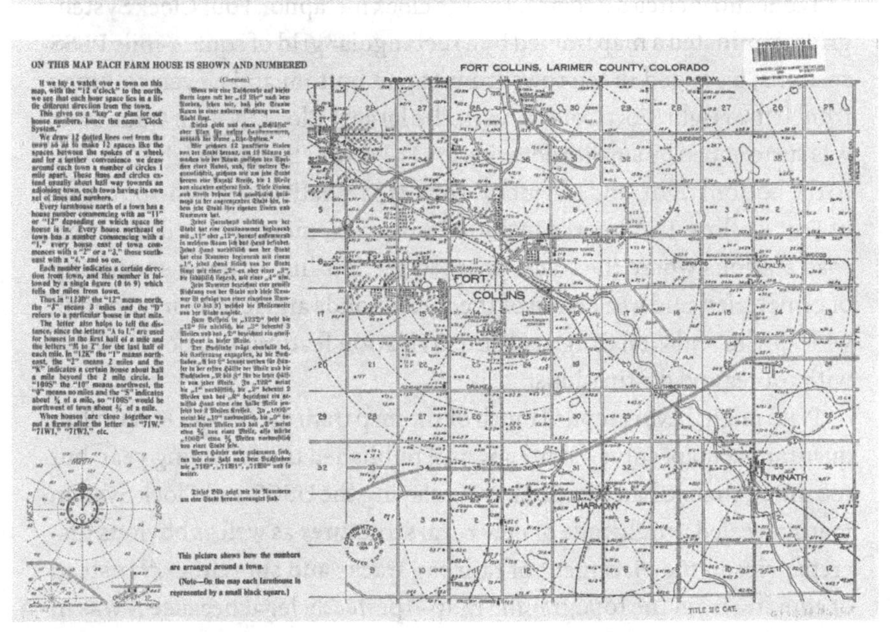

FIGURE 17. Cover (above) and map (below) of Plato's Fort Collins map, published in 1916. Original size of each: 36 × 53 cm. Scan courtesy Geography and Map Library, University of Illinois at Urbana-Champaign.

Commercial advertising was not the only messaging. Just below the title at the top of the front page, "'Clock System' Patent" proclaimed the publisher's creativity in large type. Directly below, text extending across the page declared the map "A Local Guide to Farms and Farmers," and the headings "1916 Edition" and "Revised Yearly" atop columns to the left and right envisioned longevity, updated content, and return sales. As further reinforcement, pithy paragraphs on the front and back covers offered diverse takes on the usefulness of the system, including the promise, "Every gate post with a farm number on it becomes a guide post."

On the opposite side (fig. 17) a map covering over 110 square miles was supplemented by two columns of explanation: one in English and a slightly narrower one in German, in blackletter gothic type familiar to Fort Collins's German-speaking immigrants recruited from Russia two decades earlier to harvest and process sugar beets.[43] At the bottom of the English-language column a small map key with a drawing of a watch anchoring radial lines and concentric circles underscored the clock metaphor. Four Clock System grids dominated a map framed by a rectangular grid of square-mile PLSS sections, arranged in ten rows from top to bottom and eleven columns from left to right. Fort Collins, with the largest Clock System trade area, was surrounded by smaller domains for La Porte to the upper left, Timnath to the bottom right, and Harmony, west of Timnath and south-southeast of Fort Collins. Landmarks included railroads, more than a dozen railroad stations where the tracks crossed section-line roads, a sprinkling of cemeteries, and a selection of the area's ponds and reservoirs. Though widespread and vital to agriculture, irrigation features did not interfere with routes and destinations.

Plato might have compiled much of the map from the USGS's Fort Collins quadrangle map, surveyed in 1905 and published the following year, but too much change had occurred.[44] Overlaying the USGS and Clock System maps revealed a multitude of new rural structures as well as obvious discrepancies in the placement of railroad tracks and shorelines.[45] Instead of copying from the topographic map—perfectly legal because it was in the public domain—Plato merely constructed a rectangular grid of evenly spaced section lines, drove the area's roughly 250 miles of highway or dirt roads, and plotted whatever rural residences he could spot. Because the land was rolling rather than hilly and most farmhouses were close to a road, he did not have to venture too far into a section to ask the resident's name.

Although Plato's approximate placement of farmhouses might not impress a civil engineer, greater geometric accuracy would have been irrelevant to anyone using his map and directory. A map scale approximately 25 percent larger than the USGS map's inch-to-the-mile scale accommodated the more closely spaced homes of Fort Collins's nonagricultural suburbs, whose residents might have found a Clock System address useful. More homes meant more individual sales and a larger market to attract potential advertisers.

Another information source potentially useful to Plato was a 1915 land ownership map focused on irrigated farms across northern Colorado.[46] Owners' names were added to a map of parcel boundaries that included roads, railroads, and other landscape features.

Plato might have cross-checked his list of farmers with the irrigation map, but exploring this possibility was not an option because I could not find a single copy of his directory.[47] Although the *Catalog of Copyright Entries* clearly confirms the existence of an eight-page directory (fig. 16), it somehow became separated from what seems the sole surviving copy of his Fort Collins map, in the map library at the University of Illinois at Urbana-Champaign. Moreover, neither of the two copies of the map and directory deposited when Plato registered his copyright could be found at the Library of Congress. That most of his deposit copies have apparently gone missing reflects a past policy of mismanagement and neglect.[48]

4 EXPLOITERS AND ADVOCATES

As a new map publisher with a promising product but little or no marketing experience, Plato might have had a "Now what?" moment well before delivering his Fort Collins artwork to the printer. By no means shy, he might have approached Denver-based map publisher George Clason for advice or a possible collaboration. Two years older than Plato, Clason had built a company with a dozen employees, a network of sales representatives, and the largest catalog of commercial maps west of Chicago.[1] His large and growing cartographic repertoire included a locally prominent Denver street map (fig. 6), but rural wayfinding would have been a new product line. I don't know who proposed the collaboration, but in May 1916 Clason copyrighted a "Clock system rural directory [for] Longmont, Niwot and Hygiene, Col.," a trio of growing communities between Denver and Fort Collins (fig. 13). Neither the copyright (fig. 18) nor the map mentioned Plato, linked indirectly to the project by the name "Clock System," the tiny label "Patented July 27, 1915" on the right directly below the map, and the distinctive framework of sector lines, concentric circles, and alphabetically labeled point symbols (fig. 18). Although financial arrangements or work responsibilities are unknown, Plato's eagerness to see his innovative framework put to work made him ready prey for exploitation by Clason and a few months later by a more emphatic Chicago promoter.

That the Longmont guide was not a successful collaboration is apparent in its unfinished look. The upper part of figure 18 shows a map surrounded on the left, right, and top by eleven rectangular panels. Two contain alphabetized lists of residents' names, Clock System house numbers, and post office addresses, with the A's (Abbott through Axelson) on the left and the L's (Landbolt through Lucas) on the right. The other nine panels, appar-

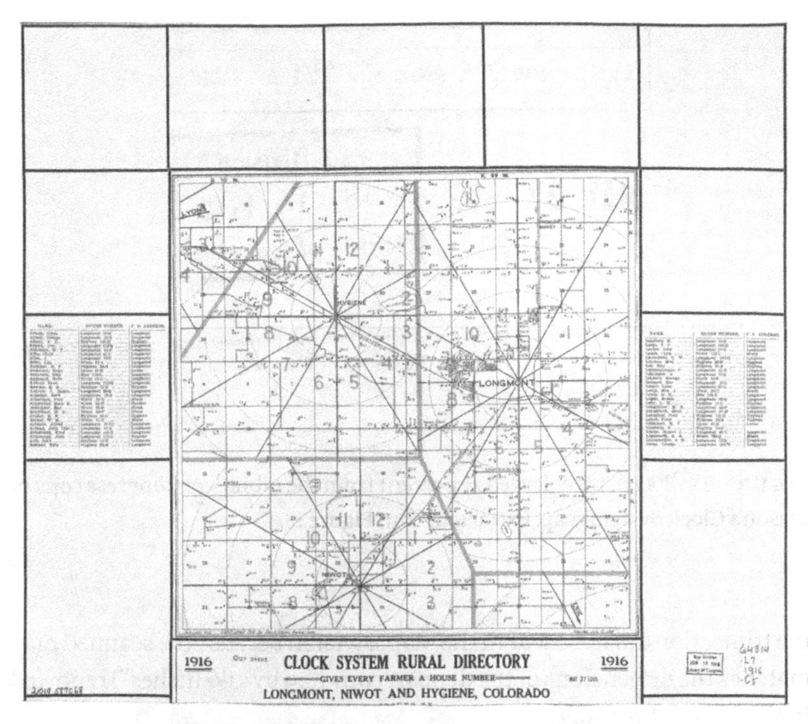

LONGMONT, Col. Clock system rural directory, Longmont, Niwot and Hygiene, Col. 1916. Size 19 by 21½ inches. [14112

© May 20, 1916; 2 c. May 27, 1916; F 28605; George S. Clason, Denver.

FIGURE 18. Clason's Clock System map in the Library of Congress (above) and its registration reported in the *Catalog of Copyright Entries* (below). Scan of map (above) courtesy Geography and Map Division, Library of Congress. Facsimile (below) from Library of Congress, Copyright Office, *Catalog of Copyright Entries*, Part 1: Books, Group 2, n.s., vol. 13, no. 6 (Washington, DC: Government Printing Office, 1916), 635.

ently intended for remaining parts of the directory, are empty. Moreover, although the map at the center has radial Clock System frameworks centered on the three places named in the title, careful inspection of the title block's bottom edge reveals that a lower portion of the map sheet might have been cut off before it was sent to the Copyright Office.[2] I am certain

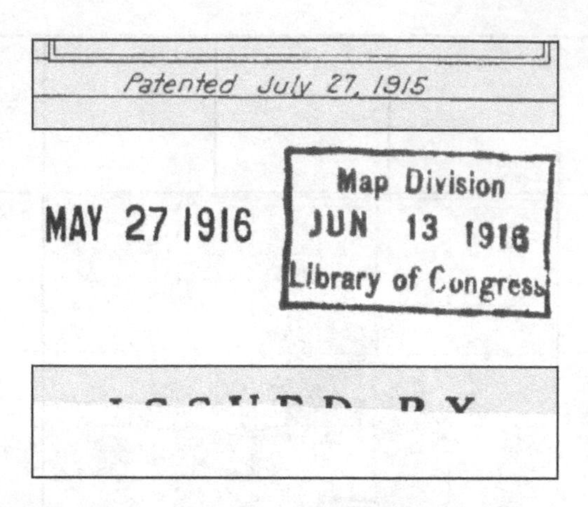

FIGURE 19. Key dates enlarged 25 percent from the Library of Congress copy of Clason's Clock System map. Extracted from Figure 18.

the truncation did not occur in the Map Division because the scanned map matches the dimensions of the deposit copy ("19 by 21½ inches") reported in the copyrights catalog.

The Longmont map was a surprise because it was copyrighted six months ahead of Plato's Fort Collins map, which I had assumed was the first published Clock System map, copyrighted in late 1916. I discovered the Longmont map by using the term "clock system" to search electronic copies of the *Catalog of Copyright Entries*—a strategy that helped me uncover many, but not all, known Clock System maps. Although the copy owned by the University of Illinois (fig. 17) confirms that the Fort Collins map had indeed been published, I have doubts about the Longmont map. Because copyright applications are not normally submitted for unpublished works, this alleged publication of an unfinished map might have been a ploy.

My suspicion is based on the tops of two words partly visible along the cut (lower edge of fig. 19). All letters are uppercase, and their juxtaposition suggests "ISSUED BY."[3] Who or what did the issuing is a mystery insofar as "issued by" is not a typical way of identifying a map's author or publisher. The wording could reflect intended free distribution by a sponsoring business or organization that backed out or was never fully sold on the idea, which might account for the map's unfinished appearance. A client's backing out could have triggered remorse or finger-pointing, and the copyright filing might have been Clason's attempt to assert ownership of his investment in drafting or photoengraving.

When the Longmont map faltered—for whatever reason—Plato surely realized that a worthwhile idea required a workable business plan. And as his Fort Collins prototype no doubt revealed, the plan's key parts included having to sell the concept to area farmers and local officials, collect farmers' names and locations, plot them on a map, assign addresses and compile a directory, manufacture and distribute mailbox signs, recruit advertisers, orchestrate typesetting and layout, arrange for engraving and printing, and determine pricing and sales. Had the Post Office been willing to revamp its rural box numbers and either pay a modest licensing fee or hire him as a consultant, he would have had few worries. Failing that, had Clason become a supportive mentor or business partner, Plato might have resisted a slick Chicago promoter's proposition that would have made him a silent partner with little or no ownership.

That promoter was Harry L. Hollister (1859–1944), who in 1917 released *Hollister's Rural Index*, a collection of classified and display ads, directory listings, and Clock System maps that "gives a number to every farmhouse in the county [by providing] an actual key to [its] geographical location [in] relation to the nearest village or city."[4] In addition to the dubious claim of "carr[ying] the endorsement of officials at the departments of Agriculture and Post Office," the *Index* was to be "a medium for advertising with great potential sales building possibilities." Clearly pitched to advertisers, the "Prospectus Sample Copy" promised an annual edition "for any County, any State."

Bold type carrying Hollister's name across the top of a title page 18 inches tall leaves little doubt about ownership (fig. 20). Look for Plato's name here or on any interior page, and you won't find it. The sole hint of his role were tiny labels noting that the "Clock System" had been patented on July 27, 1915, and that its explanatory diagram, with the iconic stem-wound watch at the center, was "Copyrighted 1916 by the U. S. R. D. Co., Colo." Never before had Plato hidden his firm's name behind a cryptic string of initials, and seldom had he missed an opportunity to repeat his folksy "lost sale" lament.

Hollister was an accomplished promoter, known for his ability to make a promising idea understandable and attractive to private investors and the public, leverage assets, and craft a plan that he could push to a successful

Prospectus Sample Copy of

Hollister's Rural Index

Opportunity Efficiency Service

THIS IS a sample of Hollister's Rural Index for any County, any State. It gives a number to every farmhouse in the county, the name of the occupant, the telephone number and the business or profession. The house number in itself is an actual key to the geographical location of each farm in its relation to the nearest village or city.

Through a "Clock System" patent—(See explanation on first page) this farm directory is made possible. It carries the endorsement of officials at the departments of Agriculture and Post Office of the United States Government because it solves a real economic need of the farmer and the Federal Government.

It is, therefore, a medium for advertising with great potential sales building possibilities.

The Hollister Rural Index
for a County is published once a year

1917 EDITION "CLOCK SYSTEM" PATENT 1917 EDITION

Section Numbers

WARNING: HOLLISTER'S RURAL INDEX is fully covered and protected by patent and copyright, and publication thereof in whole or in part by any other person, firm or corporation, is strictly prohibited. Any person, firm or corporation, issuing, selling or using any publication by whatever name, infringing such patent or copyright, will be prosecuted as provided by law. We will cheerfully pay for prompt information of any such infringement, or unauthorized use. See explanation for use on first page.

Free to Farmers and Display Advertisers whose names appear in the General Directory. To others living in the County the price is $2.00 a copy. To those outside the County, $5.00.

Published by
H. L. Hollister
155 N. Clark St.
Chicago Illinois

WANTED
An active Agent for each County in the United States. Send References.

The Number of this Book is________

FIGURE 20. Title page of Hollister's prospectus, released April 10, 1917. 30 × 46 cm. *Hollister's Rural Index* (Chicago: H. L. Hollister, 1917), i. Courtesy National Library of Agriculture.

conclusion. His greatest success was the Twin Falls irrigation and power project, which involved dams and canals, hydroelectric plants and power lines, land sales and recruitment of settlers, and financial skullduggery that enriched some investors and ruined others. Hollister's role in developing agriculture and waterpower in southern Idaho scored a privileged place in the triumphal *History of Idaho*, published in 1914, with a full-page photographic portrait highlighting his coy smile, neatly trimmed moustache, receding hairline, and dark three-piece suit. Past achievements included setting up a bank in Sioux Falls, Dakota Territory, while still in his early twenties and building the electric street railway network in Lansing, Michigan. At his headquarters in Chicago, Hollister "maintained large and handsome offices devoted to the interests of the Twin Falls country."[5] Now he was casting a larger net toward the nation's farmers.

Hollister's title page hinted at a business plan. Smaller type above the title introduced the forty-eight-page collation as a "Prospectus Sample Copy," and farther down the assertion that "The *Hollister Rural Index* for a County is published once a year" signaled a big idea with high hopes. At the lower left, a small text block revealed a pricing strategy clearly favorable to farmers, for whom the directory would be free, while other county residents would pay $2 and "those outside the county" would be charged $5. At the lower right in similar type, an appeal for "An active Agent for each County in the United States," advised "Send References."

Later that year Hollister issued a twelve-page booklet with a concise summary of his project.[6] Though *Putting the Farmer on the Map* might have helped solicit ads for the larger, folio-size prospectus, a date stamp in the USDA's copy suggests it came after, not before.[7] Although the booklet does not mention Plato by name, it acknowledges his creativity and agricultural roots: "The credit for working this out belongs to a Colorado farmer, who had both the ingenuity and bull-dog tenacity necessary for the result. He has given to the world one of the most valuable and useful ideas evolved in many a day . . ."[8]

Kind words, but instead of stopping there, the writer conflated straightforwardness with obviousness: ". . . The clock system is so simple that one wonders somebody did not think of it before, yet it can be applied to every locality, thus providing for a universal system for the United States."

Although Plato might not have appreciated the backhanded compliment, he must have been pleased by the carefully constructed paragraphs with punchy headings suggesting a comprehensive business plan. For example, "A Guide for Travelers" underscored the usefulness of placing Clock System addresses on every point on the map but also on individual signs at "every farmer's front gate."[9] The "Classified Business Directory" would simplify a farmer's shopping by identifying nearby greenhouses, dairy stock breeders, suppliers of seed corn, and various retail outlets in town, and a separate "Manufacturers' and Producers' Directory" would help him "reach the fountain head for every machine, tool or product designed for his or his family's use, or ascertain whether such advertiser has an agent in his county."[10] Of course, the *Index* would be "Free to Farmers," who could include "their telephone numbers and a three-word advertisement, all free of cost, [and because of free distribution, all readily available] to every resident farmer whose name appears."[11] And because the farmer would not discard it like an out-of-date newspaper or magazine, "its permanency and . . . daily use will make every farm family familiar with every article advertised in its pages."[12]

A daunting part of the plan was Hollister's strategy for keeping listings and advertisements current. Every county would have an "Index Advertisers' Service Bureau," to make certain advertising copy was appropriately targeted to local consumers. Each bureau would have "a salaried resident manager at the county seat and a paid representative in each township."[13] In addition to "keeping in close touch with the activities of all the people," the staff would "be on duty throughout the year and comb the business field more thoroughly than ever has been possible." Staffed by "men who are in full sympathy with the constructive purposes of the Index and who will continually revise the data," the organization "will become a free labor and information bureau to be used locally and by our advertisers." Left unsaid is a mechanism for paying the workers and coordinating their work with neighboring bureaus as well as across the nation.

Although Hollister's ambitions probably collapsed under their own weight, Plato offered a kinder explanation in his own prospectus, in the premier Clock System directory for the Town of Ulysses, New York, where he claimed, "In 1916 and again in 1919 the inventor gave some Chicago men an opportunity to develop the patent but owing to the war their plans could not be carried out, and their connection with the work now has entirely ceased."[14] The plural "men" points to the involvement of Andrew B. Hulit (1866–1938), a Hollister lieutenant whose own exploits include a 1912 plan to build a $5 million agricultural exhibition hall in Chicago, for which Hulit falsely claimed several "Big Men" as backers.[15] His taste for huge budgets reemerged on April 18, 1917, when the (Madison) *Wisconsin State Journal* identified him as the "manager" of a "nation-wide survey of agricultural interests" purported to cost $25 million and "financed by [H. L.] Hollister, who owns the index system."[16] Hulit was in Madison "to confer with state and university officials" about "get[ting] students at universities to direct the survey in [the] counties." Nothing was said about Plato, but the inventor's contribution was apparent in Hulit's claim that "the Survey will result in every farmer in the United States receiving an address such as has every residence in cities."

Three weeks later, the Stevens Point (Wisconsin) *Gazette* identified Plato as the inventor and cited his "lost sale" as the inspiration. Discussions with university officials had apparently been successful because the project now included "a competent college man being employed as canvasser to visit every farm."[17] And although Hollister and Hulit "had planned to start the system in Illinois," they had been "induced to try it out first in Wisconsin."

Selection of Wisconsin for their pilot surely reflects the enthusiasm of Charles Josiah Galpin (1864–1947), associate professor of agricultural economics at the University of Wisconsin and director of its research program on rural life and work. Galpin had started his academic career as a schoolteacher but drifted into what became the academic discipline of rural sociology in his late forties after going to Madison (where the university was located) in 1911 and using field surveys and graphic analysis as research tools.[18] To emphasize the importance of giving the farmer "a business man's address . . . with reference to his business center,"[19] his seminal 1918 book *Rural Life* not only mentioned Plato's Fort Collins map but retold the "lost sale" yarn:

> A Colorado farmer in the pure-bred Guernsey business, pushed by the necessity of advertising his pure-bred calves in the Fort Collins newspapers, like any other Fort Collins business man, has devised an ingenious method for giving every farmer such an address. In fact his method has been patented, and a rural directory has been published and copyrighted, based upon the method. The accompanying chart of an actual rural community illustrates the so-called "clock system" of this Colorado farmer in operation.[20]

Although Galpin omitted the inventor's name and confused the timeline—the lost sale would have occurred at Semper at least a year before Plato's Fort Collins work—he demonstrated an appreciation of the idea by including his own version, captioned "A Rural Directory by the Clock System."

Figure 21 illustrates the structure and details of Galpin's map, which he reproduced at a smaller scale than Plato typically used. To better fit the

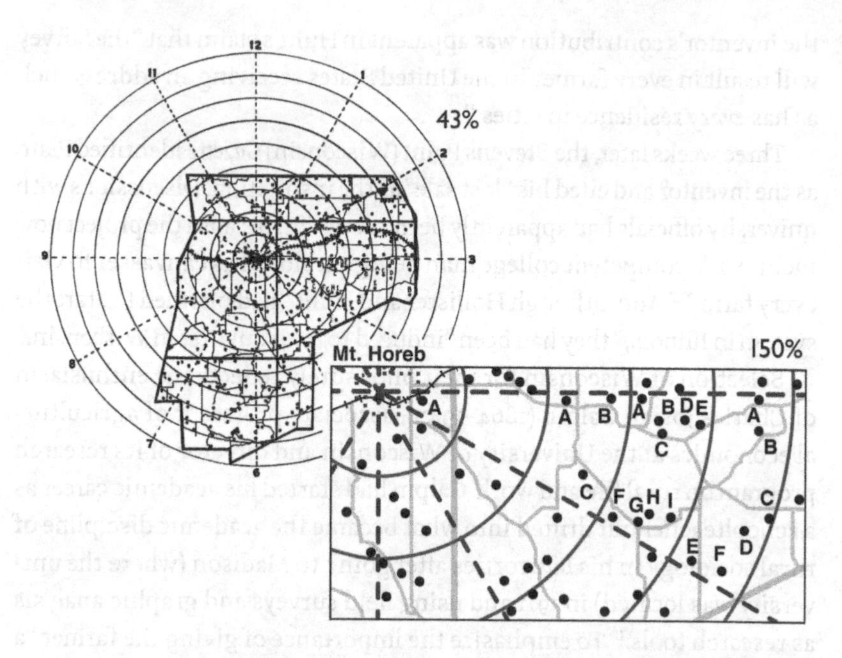

FIGURE 21. Galpin's map of the trade area of Mt. Horeb, Wisconsin, was based on Clock System principles. Enlarged and redrafted portion of the 3 o'clock sector (lower right) complements the 43 percent reduction of the image in Galpin's book. 13.5 × 13.5 cm. Adjacent circles are a mile apart. Reduced from scan of map in Charles J. Galpin, *Rural Life* (New York: Century, 1918), 344. Lower-right portion redrafted by the author.

page, my image is smaller still, at 43 percent of Galpin's scale. Even so, it captures the fundamentals of the twelve sectors and their division into neighborhoods called "blocks," in this case, by nine concentric circles a mile apart and centered on the village of Mt. Horeb, approximately 15 miles west southwest of Madison. The heavy line represents the trade divide between Mt. Horeb and surrounding village centers: within this boundary farm families generally had more contact with businesses within Mt. Horeb than with providers in nearby villages.[21]

To compensate for the further degradation of Galpin's poor inkwork, I redrafted the first five blocks in the 3 o'clock sector (fig. 21, lower right)—the only sector for which Galpin labeled dots and explained the addressing scheme—at a size one and a half times the scale in Galpin's book. For completeness I included portions of surrounding neighborhoods, and for clarity I adjusted gray levels to make the dots and their labels stand out. My enlarged version makes it possible to appreciate Galpin's claim that a map of rural business addresses made it "as easy for the farmer to advertise his goods as [it was] for a merchant [in town]."[22] Moreover, by demonstrating the existence of rural trade areas, maps like this were useful in urging postal officials to reconfigure delivery routes "with conscious reference to . . . comprehensive business centers."[23]

Galpin was not fully convinced that Plato's addressing scheme was the best way to give farmers a business address. After all, rural roads could be named and farmsteads numbered, as they are today, in accord with the E-911 system for dispatching emergency vehicles; in a pre-GPS era, street and address signs could make this strategy work as long as maps provided an overview. Nonetheless, farmers were beginning to recognize the advantages of legible (and perhaps even artistic) signboards. Moreover, "the movement for distinctive farm names indicates that farmers are getting ready for business advertising."[24] Perhaps Galpin had heard of Featherton Farm.

Plato found another academic advocate in Dwight Sanderson (1874–1944), hired by Cornell University in 1918 to set up the new Department of Rural Social Organization in the College of Agriculture. Sanderson had recently stepped back from successful careers in applied entomology and academic administration to undertake a doctorate in sociology at the University of Chicago, which awarded him a PhD in 1921.[25] His dissertation "The Rural Community: A Social Unit" and the many research projects he directed during his quarter century at Cornell reflected a deep interest in mapping, central locations, and community areas. In June 1920, in one of his first Extension Service reports at Cornell, he not only commented approvingly on Plato's invention but also devoted five pages and four illustrations to the Clock System. In addition to including the apparently irresistible "lost sale" story, Sanderson applauded "numbering farms by community" as an antidote to the farm home's lack of any "designation by which it can be easily located."[26]

Unlike Galpin, who denounced the RFD number as an unworkable business address, Sanderson saw a growing problem of community identity. Cost-cutting by the Post Office Department had closed many rural post offices and assigned their rural-delivery routes to "some railroad station or larger town which [the farmer] visits only occasionally"—unfortunate because "the outside world knows him only by his post-office address and may never hear of the community in which he really lives."[27] Were postal officials to adopt Plato's system, "the identity of the community would be recognized and community consciousness would be promoted."

The academic and the inventor shared a common interest. Sanderson collected data on farmers' social and business networks and used these data to study central places and rural communities, and Plato needed a workably small number of address hubs for his maps. Like Galpin, Sanderson recognized that trade-area boundaries were momentary compromises sensitive to improved rural roads, growing automobile ownership, and farm consolidation and abandonment. Because farmers could be connected to different places in diverse ways, he looked at the number and type of services provided in a village, including schools, churches, social foci such as a grange hall, and economic institutions such as a collection station for a regional milk-marketing cooperative. A small rural community based largely on where people did their basic shopping or attended church could still have social significance, even if it was part of a larger community with higher-level attractions or was split among two or more larger communities. After all, shopping for furniture or major appliances in a city like Ithaca did not negate a farmer's relationship to a much smaller locality. Sanderson and Plato needed to identify these basic rural localities.

Sanderson explained these relationships with a map for Caroline Township, southeast of Ithaca but still within Tompkins County (fig. 22).[28] The township had five villages: Slaterville Springs, Caroline, Caroline Center, Brookton, and Speedsville. Slaterville Springs, in the north-central part of the township, had two physicians, two blacksmiths, a hotel, a tearoom, and a post office, as well as the Dairymen's League, the grange, the Masonic Lodge, and the Red Cross, not to mention a store and two churches. By contrast, Caroline and Caroline Center, which had little more than a church and a school, were almost wholly within the "grange area" and "milk area" boundaries of Slaterville Springs, a somewhat larger and clearly more self-sufficient community. Even so, farmers in the neighborhoods of Caroline and Caroline Center continued to patronize their local store, and community boundaries on Sanderson's map attested to these village's usefulness as address hubs.

So, did Plato collaborate with Sanderson or did they use different criteria in delineating community areas? Figure 23 shows general agreement as

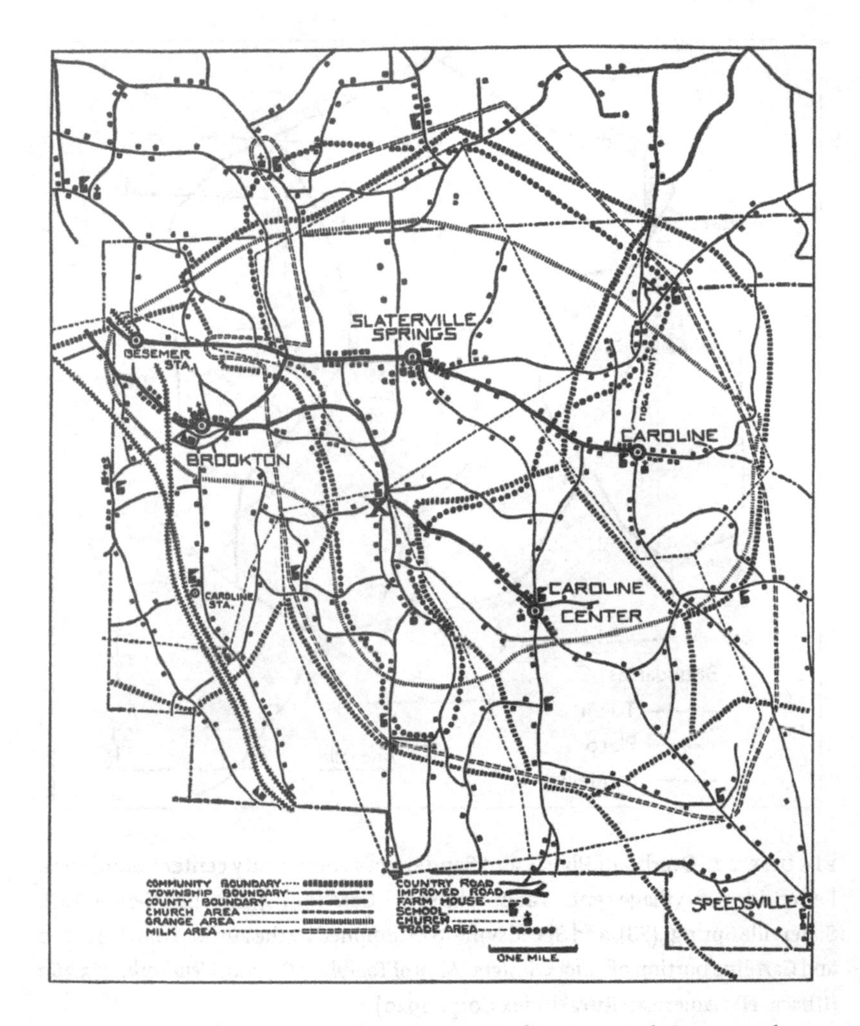

FIGURE 22. Sanderson's map of Caroline Township, New York, juxtaposed community boundaries for all five villages with grange-area and milk-area boundaries for Slaterville Springs. Area shown measures 8.7 miles (14 km) from left to right. 10.6 × 12.9 cm. Dwight Sanderson, "Locating the Rural Community," *Cornell Reading Course for the Farm*, lesson 158 (June 1920): 415–36, fig. 144, on 423.

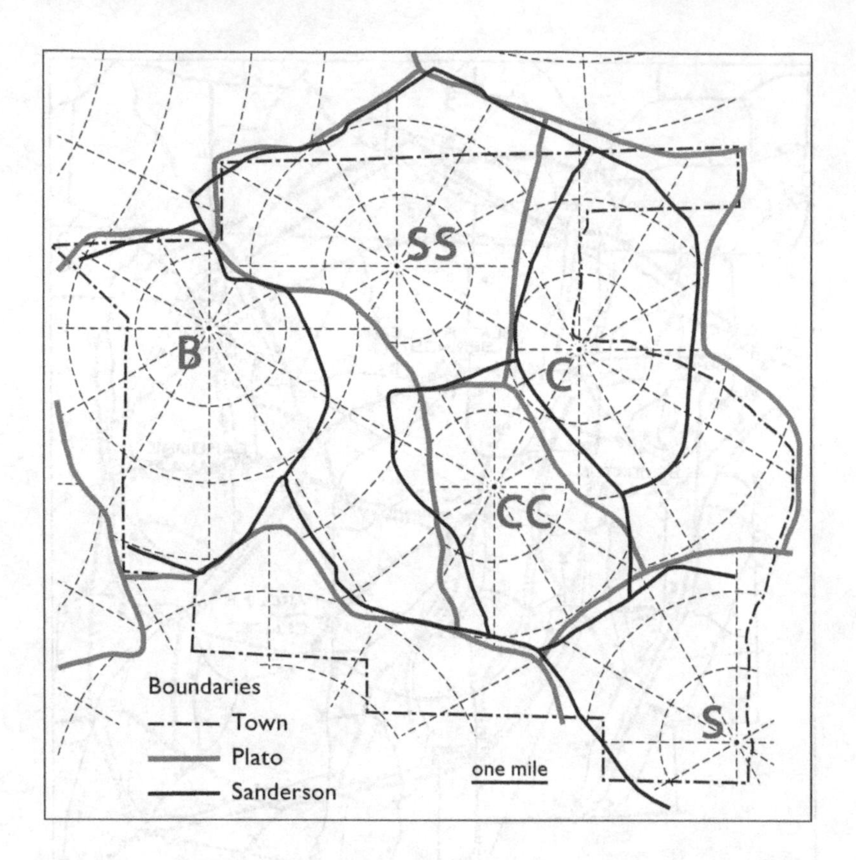

FIGURE 23. Overlay of Plato's and Sanderson's community center boundaries. Letters identify village centers at Brookton (B), Caroline (C), Caroline Center (CC), Slaterville Springs (SS), and Speedsville (S). Compiled by the author from Figure 22 and Caroline portion of *"Clock System" Map of Tompkins County, New York*, 1:63,360 (Ithaca, NY: American Rural Index Corp., 1920).

well as substantial differences between their maps, both published in 1920. I constructed this cartographic comparison by tracing sector lines, circles, and address-hub and town boundaries from Plato's map of Tompkins County, by tracing village centers and community-area and town boundaries from Sanderson's map of Caroline, and by making minor adjustments so that Sanderson's centers aligned with Plato's.[29] These adjustments did not obscure the facts that the two maps recognize the same five village centers and that their community-area boundaries are generally but not precisely similar.

Disparities suggest simple explanations. The strongest agreement is apparent around Caroline Center (CC), where boundaries rarely more than a half-mile apart affect relatively few farmsteads. By contrast, although the boundaries surrounding Brookton (B) and Slaterville Springs (SS) are generally similar along their northern and southern reaches, they differ markedly over a tongue of land extending five miles southeastward from Brookton—Sanderson considered this a salient tributary to Slaterville Springs, whereas Plato assigned it to Brookton (thick gray line in fig. 23), probably because the improved highway into this area (thick black line in fig. 22) is aligned to Brookton. Perhaps the farmers interviewed by Sanderson allowed their milk and grange connections to influence their everyday shopping.

Overall, Plato's divides seem more influenced by road distance and highway conditions than Sanderson's. Where the rural sociologist relied upon a carefully structured field survey, the Clock System guru consulted the county agent. Differences on the eastern side of the village of Caroline (C) probably reflect the county agent's reluctance to assign farmers to a village outside Tompkins County. Boundaries generally concur for Speedsville (S) in the extreme southeastern corner of the town.

Although Plato and Sanderson no doubt knew each other, and perhaps even socialized—Sanderson was just two years older and had moved to Ithaca the same year—Plato's relocation in 1918 more likely reflected his attendance at the Cornell Winter Course in early 1896 and encouragement by Liberty Hyde Bailey (1858–1954), founder of Cornell's College of Agriculture, and Albert R. Mann (1880–1947), the college's current dean, both of whom strongly supported rural sociology and community identity. His move might also have been encouraged by Galpin, who left Wisconsin in 1919 to direct rural life studies at the USDA; Charles A. Lory (1872–1969), president of the Colorado Agricultural College; and Eugene C. Branson (1861–1933), chair of the Department of Rural Economics at the University of North Carolina. Plato acknowledged all of them in his Ulysses directory.[30]

5 ITHACA

Ithaca, the county seat of Tompkins County, is surrounded by nine rural towns. In late 1919, a little more than a year after his return, Plato published his second Clock System map, focused on the Town of Ulysses, several miles to the northwest on the shore of Cayuga Lake. The map occupies a single 9 × 12-inch page in a thirty-two-page directory bound in an unpaginated cover printed on thick paper (fig. 24).[1] Six mostly shorter directories followed in late 1920, with each covering one or a pair of the county's other rural towns. Plato ignored the City of Ithaca, already covered by a traditional city directory with a map on the inside back cover.[2] His composite Clock System map of Tompkins County, published in 1924, largely ignored locations within the City of Ithaca, which was surrounded by the Town of Ithaca, a separate political unit. The directory omitted residences within the city proper, and the map showed only the more important city streets. Even so, Ithaca was an address-hub for areas outside the municipal boundary but not within the trade area of one of the surrounding business centers.[3]

Although focused on a single township, the Ulysses directory was a prospectus crafted to introduce the Clock System to farmers and advertisers throughout New York State. Interwoven with its alphabetic and numerical lists of addresses were richly illustrated multipage articles titled "Putting the Farmer on the Map," "Explanation of the Clock-System," and "The First Rural Index." A "Special Notice" addressed to "national and state agricultural officials, county agents, school officials and teachers, newspapers and advertising agents," touted the project as "home grown" and promised that "a number plate [would be] placed on the road in front of every farmhouse without cost to the farmer, [who would] also [be] given

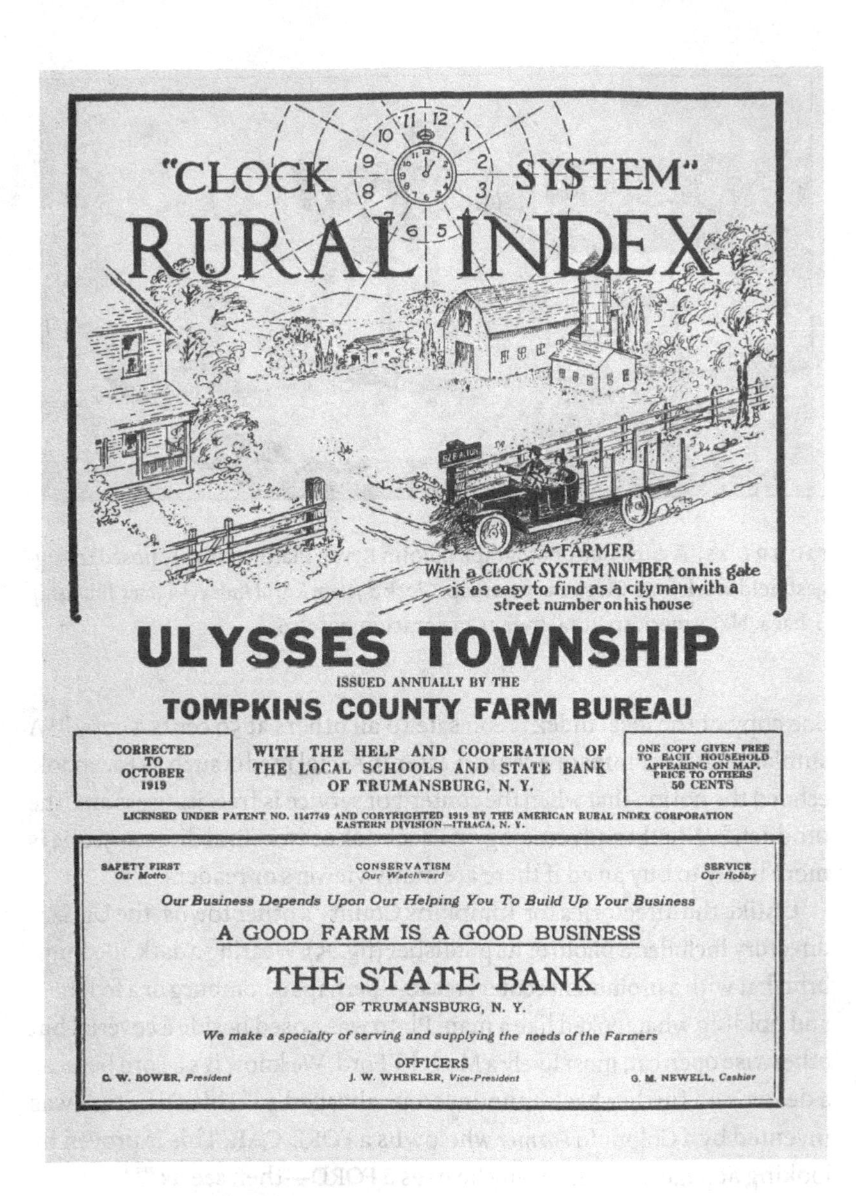

FIGURE 24. Cover of the Ulysses Township directory (1919). *Clock System Rural Index, Ulysses Township* (Ithaca, NY: American Rural Index Corporation, 1919), front cover.

FIGURE 25. A circa 1918 photograph of John Byron Plato, probably posed to suggest field work for a Clock System map. *Clock System Rural Index, Ulysses Township* (Ithaca, NY: American Rural Index Corporation, 1919), 6.

one copy of the local index . . . on sale to all others at 50 cents a copy."[4] A similar strategy, rampant a century later in social media such as Facebook, echoed the notion that when the content or service is free, its users are "the product."[5] Whether advertising on Facebook or in a rural directory, one is more likely to buy an ad if there are many viewers or readers.

Unlike the directories for Tompkins County's other towns, the Ulysses directory included a photo of its publisher (fig. 25). Wearing a dark, medium-brim hat with a prominent center crease—perhaps a homburg or a fedora—and holding what looked like a map, Plato was posed beside a covered but otherwise open car, most likely a Model T Ford. We know it's a Ford because a dealer's ad further back announced unabashedly, "This Directory was invented by a Colorado *Farmer* who owns a FORD CAR. This is proven by looking at page 6. Ask him why he uses a FORD—then see us."[6]

In addition to flattering a local business, the ad affirmed the inventor's empathy for farmers, who were repeatedly reminded that Plato was one of them. The inside front cover carried an endorsement by Liberty Hyde Bailey, former dean of Cornell University's agricultural division, who not only asserted that a farmer's home "should be a place on the map and a recognized unit in the community" but also recalled that Plato "was my student more than twenty years ago."[7]

The caption below the photo quoted the inventor's recollection, "Milking my 10 cows every day gave me lots of time to think how much better it would be if we farmers had a real address."[8] Putting farmers on the map was advantageous because, as another article in the directory argued, "A definite address has a big cash value to a farmer. Suppose you need a veterinary or doctor in a hurry and you cannot get your regular doctor. Whoever you are able to get will lose valuable time trying to find your place unless the community has been put on the map."[9]

Further on, under the clichéd heading "Necessity the Mother of Invention," the same article noted, "The patented plan used in this Rural Index was developed in Colorado by a farmer, J. B. Plato, who tried to get people to come and see some pure-bred Guernsey calves that he had for sale. It was while trying to write an advertisement to put in his local paper that it was forced upon him that he was absolutely without any form of real address."[10]

As Plato's first cartographic publication in Ithaca, the *Clock System Rural Index, Ulysses Township* served as a prospectus for later guides and affirmed his ties with the Cornell community. Shortly after his return, the *Cornell Countryman*, a monthly magazine produced by students in the College of Agriculture, published his article "Numbering Farm Houses" in its April 1918 issue and identified him in the byline as "J. B. Plato, '96"—which conflates his eleven-week "Short Winter Course" for farm youth with a four-year degree.[11]

The earliest evidence of Plato's move to Ithaca is his draft board registration card, dated September 12, 1918, almost a year and a half after the United States declared war and two months before Germany's surrender.[12] A native-born, forty-one-year-old white male with medium height and build, gray eyes, and brown hair, Plato was employed as a machinist at the Thomas-Morse Aircraft Company, manufacturer of the S-4 biplane, used for training pilots.[13] His education at Manual High and experience making horse-hitches qualified him for paid employment at Thomas-Morse while he set up his mapping business. He had a permanent residence at 319 East Mill Street, where his mother, Helen Plato, would join him before the federal census taker visited in late January 1920. To affirm the permanence of his move, Plato surrendered his Semper Garden Tracts lots by defaulting on tax and mortgage payments. Apparently unwilling to sever ties to Semper and his iconic status as a Colorado farmer, he hung onto the recently purchased 0.4-acre lot near the Denver and Interurban trolley stop.[14] Helen

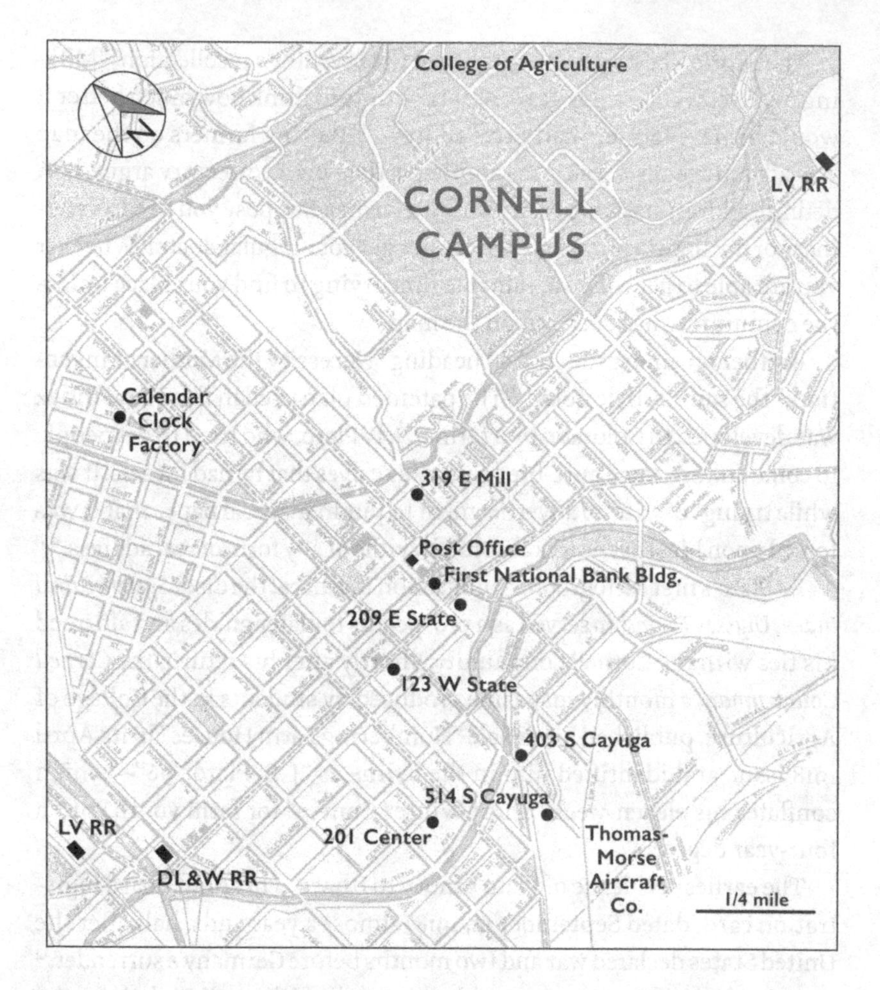

FIGURE 26. Locations relevant to Plato's work and residence in Ithaca, New York, 1918–31. Stations of the Delaware Lackawanna and Western (DL&W) and Lehigh Valley (LV) railroads connected Ithaca with New York and Buffalo, and a branch line through the Lehigh Valley's East Ithaca station, on the right near the Cornell Campus, provided less direct connections with Syracuse and the New York Central's mainline. Elevations at the top of the map were nearly 500 feet above areas at the bottom left. Compiled by the author from city directories and Sanborn Map Company, *Insurance Maps of Ithaca, Tompkins County, New York* (New York: Sanborn Map Co., 1919). Plotted on portion of *City of Ithaca, New York* [map] (Ithaca: Ithaca Chamber of Commerce, 1947).

had already left Denver and in early 1918 was living with relatives back in Geneva, Illinois.[15]

A short walk from downtown Ithaca (fig. 26), 319 East Mill Street was near the bottom of a long, steep hill separating Cornell's central campus (including the College of Agriculture) from the lower, comparatively level part of Ithaca known as The Flats. The 1920 Census listed two heads of household at 319: Sarah Aileen Brown (fifty-nine, single), the owner, and Plato, a renter. The number "1" next to his name suggests he and his seventy-seven-year-old mother, possibly averse to climbing stairs, had a first-floor apartment, whereas Brown, with a "2" next to her name, might have had the second floor for her household, which included three other occupants, each identified as a "lodger." Several other buildings in the neighborhood had lodgers, and two had live-in maids. Neighbors' occupations included chauffeur, laborer, store clerk, plumber, electrician, and optician. Plato was no longer working at the aircraft plant; the enumerator left his place of work blank and reported his occupation as "farmer."

Calling himself a farmer could not disguise Plato's new role as a promoter intent on leveraging enthusiasm for his brand, marshalling local support, and applying what he picked up from watching Hollister and Hulit. To call his American Rural Index endeavor a "Corporation" required recruiting a board of directors and getting certified.[16] On March 1, 1919, the *Ithaca Journal* reported that New York's secretary of state had chartered the company to raise $10,000 for a "directory and advertising business in rural communities."[17] Plato was a director along with Charles E. Treman (Cornell '89 and president of the Ithaca Trust Co.), Juan E. Reyna (Cornell '98 and CU professor of agricultural engineering), Charles Tracey Stagg (Cornell '02 and a local attorney and member of the CU law faculty), F. H. Springer (an executive at the First National Bank), H. J. Van Valkenburgh (president of Ithaca Engraving Co.), and John W. Baker (vice president and manager of the *Ithaca Journal*).[18] With a host of distinguished backers in place, Plato was now free to work with his key partner, the Tompkins County Farm Bureau.

Self-identification as a farmer might have helped sell the Clock System to the Bureau, which signed on as copublisher of the Ulysses directory and six separate directories, each covering one or two other townships, as well as a combined directory issued in 1920 for the entire county. As an advocacy, lobbying, and educational organization—the agricultural equivalent of a chamber of commerce—the Bureau's stake and role in the directories were described in a pair of concisely crafted news articles that resembled press releases.[19] The organization had not only assumed responsibility for the metal address signs affixed to each farm's roadside gatepost but was also coordinating the participation of local schools, the College of Agriculture, and local business groups.

Plato now had the main elements of a simple business model.[20] *Key partnerships* included the Tompkins County Farm Bureau, Cornell's College of Agriculture, the county agent, and local school systems, which were responsible for collecting data within their districts. Although *key resources* included county extension agents and Cornell's rural sociologists, the primary resource was Plato's Clock System patent, which as the business's *brand* was tied to the *value propositions* of giving farmers a more useful business address and reducing rural isolation. Farmers and businesses selling to farmers were his *customer segments*; advertising and sales of maps and directories were his *revenue streams*; and newspaper stories, individual correspondence, and Plato's involvement in civic activities loosely constituted the firm's *customer relationships*, which were inherently indirect because the plan did not include door-to-door sales or walk-in customers.

Plato strongly believed in the efficiencies and cost savings only possible when the work "is divided up among those who can each do their own part best."[21] A strategy built around an Ithaca-based "central organization" consisted of six steps, which could be implemented in any rural county:

Step One—Securing exclusive rights, complete instructions and necessary supplies from the central organization.
Step Two—Having each rural school of the township collect the data for its own school district.
Step Three—Having the high school and farm bureau collect the data from each rural school.
Step Four—Having the central organization prepare a map and map plate ready for printing. This map shows the location of all roads and farmhouses, and beside each house on the map is its proper number.
Step Five—Having the local newspaper solicit enough local advertising from farmers and merchants to pay the local cost of printing a "Rural Index."

Step Six—Having some one [sic] paid to place the proper number plates on the road in front of each farmhouse and to deliver the books.[22]

Outsourcing was a guiding principle. The local newspaper in Step Five might be a weekly that served one or more towns, and installation of the metal number plates could be farmed out to local youth. As Plato noted later in the Ulysses directory, "High school boys or others will be glad to put up these signs for from 5 to 10 cents each."[23]

Like any practicable strategy, Plato's plan included relevant details. To help pupils and teachers collect data, the central organization would provide blueprint maps based on USGS topographic maps but drawn at three inches to the mile.[24] Because these maps were published at only one inch to the mile (1:63,360)—too small for convenient annotation—he enlarged them with a pantograph (fig. 27), a mechanical device for copying a drawing at a smaller or larger scale.[25] Plato had the skill needed to produce an enlarged pencil drawing, readily duplicated at a local blueprint shop. USGS maps showed landmark buildings, such as farmhouses and sometimes large barns, but not sheds and chicken coops. These symbols told the field mapper where to visit and were more exact than positions added merely by eyeball. A 3:1 enlargement allowed sufficient room for writing the resident's name next to a farmstead's dot, and additions or deletions could be noted on the blueprint.

Elementary school districts in rural New York were smaller than the attendance zones of today's consolidated schools, and their vague boundaries required the teacher or principal to "draw on their map the outline of their school district as they understand it."[26] These maps were then collected at a nearby high school, where "some of the advanced pupils" would cut and paste together a "complete and corrected township map." The county agent would then "select the natural centers from which the clock numbers will radiate and . . . mark the boundaries between these centers"—fully consistent with the principle of letting each contributor "do their own part best."

Thinking ahead to rural directories for counties well beyond Ithaca, Plato delegated selection of a local printer to the county agent, and let the printer solicit local advertising, set rates, and orchestrate layout. The county agent, who had to approve the layout, could "also reserve such space as he

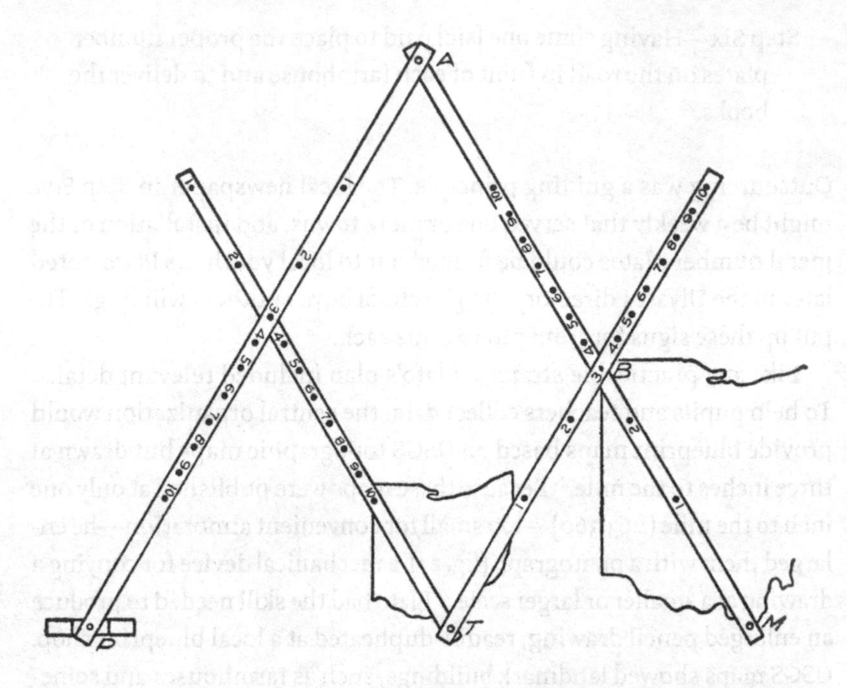

FIGURE 27. Principle of the pantograph, as described in a 1918 textbook on engineering drawing. The device consisted of four wooden or metal strips connected to form a parallelogram and anchored to a drafting table at a single pivot point (P). Moving the tracing point (T) along a feature on the original drawing caused a pen or pencil at the marking point (M) to reproduce its shape at a larger size. The strips were joined permanently at A and T, and at two movable points by inserting pins in numbered sockets configured to ensure a parallelogram. The number of scale changes was fixed, and the strips had to be connected by inserting pins in sockets with the same number (three in this example). To reduce the scale, the draftsman swapped the tracing and marking instruments and placed the original drawing on the right. Thomas E. French, *A Manual of Engineering Drawing for Students and Draftsmen*, 2nd ed. (New York: McGraw-Hill Book Company, 1918), fig. 546 on 303.

may want for his own use." All of this would be printed on the opposite side of a Clock System map sheet provided by the central organization, which controlled design and content. Local pupils would scrutinize the map and compile an alphabetical list of farmers, their addresses, and perhaps a phone number, along with a specialization such as dairy or "pure-bread Guernseys," and the list would be typeset and printed locally. Lest this decentralized collaboration seem too daunting, Plato offered the confident reassurance of a trusting, team-building promoter: "There is nothing at all difficult about any part of this work and no teacher or county agent need feel any hesitation towards doing their share of the project, which when once established will become a lasting benefit to the whole community."[27]

Topping off Plato's business plan was a pricing policy reminiscent of Hollister's free-to-farmers mantra. Farmers would receive a free copy of the map and directory as well as a free address sign for their gatepost.[28] Underscoring the benefit to the whole community, Plato pledged not to accept advertising from mail-order firms in likely competition with local businesses. Although a local printer or other local organization would receive income from local advertising, central operations would be supported by ads for manufacturers of farm and household supplies sold by local retailers as well as by the sale of individual copies, priced at only fifty cents and sold at local groceries, druggists, banks, and garages to promote a circulation wider than just farmers, and thus offer advertisers a broader marketplace.

A final component was the directory's cover, designed and produced by the central organization and printed on more durable colored stock. As exemplified by the Ulysses directory (fig. 24), the cover carried an ad for a local bank, which was guaranteed "exclusive use of the front cover of all the books issued during the first three years," and which paid extra for the "privilege of sending out all the farmers' copies during this time"[29]—if the directory came from a bank, wouldn't it be seen as a gift from the bank?

The "first three years" guarantee presupposed regular updates, described as a straightforward, low-cost process. The post office and schoolchildren could update the list of residents, and because Clock System addresses did not change, a new rural residence could be added by going a step further down the alphabet. Unless damaged, address signs were not replaced. Advertisers could be charged a new fee, and the central office could focus on expanding to additional counties. Reinforcing the notion that a rural index was self-supporting, the farm bureau could claim (as the Tompkins County Farm Bureau did in its page-one pitch "To the Farmers of Ulysses") that the plan was "simple, convenient and workable" and the Bureau incurred no added expense.[30]

The final element in Plato's business plan, its *cost structure*, includes a large amount of volunteered time: economists label this an externalized cost, meaning that someone else pays for it. Plato would have drawn a salary, and the central organization had to pay for paper and printing, buy furniture and stamps, pay for topographic maps and blueprinting, and rent workspace, though the lack of an address other than Ithaca, New York, suggests that during the business's first few years, space might have been donated, along with the services of the county agent, who arguably was doing his job. Teachers were also doing their jobs when the cost of data collection was shifted to schoolchildren, whose participation was touted as a learning experience. The Ulysses directory included a picture of a school district map "drawn by a 9 year-old girl," and its caption quoted pupils as saying "making a map is fun" and their teacher applauding this "fine lesson in geography."[31]

Clearly, Clock System cartography required considerable good will, much of it sustained by local newspapers in Ithaca and elsewhere, which agreed that RFD route numbers contributed little to the farmer's sense of place. By contrast, Plato's "lost sale" narrative and his notion of a "real address" had a distinct appeal, as did sample Clock System maps, sometimes reproduced full-size, as the *Ithaca Journal* did in its January 18, 1919, issue which announced the premier map, for Ulysses Township, and credited "Professor Olney's class" at Trumansburg High School for "charting the town."[32] In addition to acknowledging the Farm Bureau's role as publisher, the article attributed the "unique plan" to Plato, "who formerly lived in Colorado, but who is now living in this county."

Plato had become a community asset. A February 1920 story headlined "Ithacan's Invention to Put Farmer on the Map Finds Place in Digest" noted with obvious pride a two-page *Literary Digest* article (which read like an unabashed press release).[33] Fifteen months later the *Journal* credited the *Digest* piece for "offers and requests from more than 30 states" as well as the possibility that "the local concern which started in a small way may develop into a large map publishing business."[34] On a triumphant note, the story praised "the Ithaca men who have carried on this work throughout the trials of experimental development."

Plato's own activities appeared occasionally in the local news. A September 1919 ad for the Tompkins County Farmers Co. listed him among several dozen prize winners at the Trumansburg Exposition: the Tompkins County equivalent of a county fair. Along with four other fairgoers, he had received a "100 Unicorn," most likely a premium sponsored by a seed company and awarded in a drawing.[35] A reporter alert to national precedents found one the following August, when Plato explained the Clock System at an exhibit hailed as "a feature never before shown at any fair."[36] Although he wasn't named in a story reporting record attendance the following year, twenty visitors to the *Journal*'s tent were identified by their name and Clock System address.[37] Plato's own name came up occasionally in the newspaper's "Social and Personal" column, which listed him among various Ithacans staying at a New York hotel or noted his "spending a few days in Springfield, Mass."[38] In late August 1920, the *Journal* reported that his mother, "who has been living with [him] on East Mill street for the past year, has left for Geneva, Ill., where she will remain during the winter."[39] Ithaca had a small-town ambiance with little difference between nosey and newsy.

Plato was actively involved in several local service and business organizations. In spring 1921 he and two other men organized a new Boy Scout troop, which was officially recognized by national headquarters the following December.[40] In May 1924 he was a founding member of the Advertising Club of Ithaca, which a year later received its official charter from the Associated Advertising Clubs of the World.[41] The following winter he helped plan and judge a children's ice-skating contest.[42]

Positive coverage by the *Ithaca Journal* might reflect Plato's connection with Frank Gannett (1876–1957), a Cornell alumnus who was part owner of the *Journal* and several other Upstate New York dailies, which tracked expansion of Clock System mapping into their own counties.[43] Although Gannett had been a student at Cornell when Plato was on campus in 1896 for the eleven-week Winter Course, it is highly unlikely that they knew each other before the inventor returned to Ithaca. Although Plato was not hesitant in claiming a Cornell affiliation—recall that the byline for his 1918 article "Numbering Farm Houses" in the student publication *Cornell Countryman* read "J. B. Plato, '96."[44]—the 1920 *Cornell Alumni News* article that identified several university alumni as founding members of the American Rural Index Corporation board described him only as "Mr. J. B. Plato, formerly of Colorado, now of Ithaca."[45] And the single-paragraph January 1922 *Alumni News* item reporting that "Frank E. Gannett '98" had been elected president of the directory company did not mention Plato at all.[46] How the inventor and the news publisher connected is unclear insofar as Gannett was living in Rochester.

The more expansive *Journal* article reporting Gannett's election mentioned that the directory company "will move into new and more commodious quarters" but did not mention a location.[47] The 1923 city directory and a legal notice announcing the next shareholders meeting, scheduled for December 6, 1922, confirmed that Plato's firm had moved to a factory building that had housed the Ithaca Calendar Clock Company (fig. 26), which went out of business in 1918. As the name implies, a calendar clock has two dials: a clockface above and a dial with month and date hands below. The legal notice for the previous meeting suggests that Plato's firm had been sharing the quarters of the Ithaca Engraving Company, on the fourth floor of the First National Bank Building.[48] Moreover, a July 1920 news story revealed an earlier downtown location (fig. 26) by advising teachers eager to map their school district "to get a blue print from the office of the American Rural Index Corporation, 123 West State Street by writing or by calling at the office"—also the home of the *Ithaca Journal*.[49]

Figure 26 illustrates key distances and proximities for Plato's decade and a half in Ithaca. Rotating the base map forty-five degrees counterclockwise provided an efficient configuration framed at the top by Cornell's College of Agriculture, at the bottom by the Thomas-Morse Aircraft Co. plant, and on the left and right by the city's three railway stations.[50] Faintly visible in the background are several streams that flow westward down a steep slope and account for the slogan "Ithaca is gorges." Cornell's campus typifies the higher ground preferred by higher education; the walk uphill from Plato's first residence at 319 East Mill Street (now East Court Street) would have been more exhausting than a one-mile stroll across level ground.

Plato occupied the East Mill Street apartment for less than five years. City directories and the 1925 Census for New York State show relocations to 201 Center Street (1923), 514 South Cayuga Street (1925 Census), 403 South Cayuga (1925), and back to 201 Center (1927). All were rooming houses in less attractive and less expensive neighborhoods, and all were a short walk to downtown.[51] In these accommodations Plato and his mother were roomers ("r"), except for 1925 through 1927, when the directories listed his mother as a boarder ("b"), implying that she regularly ate with the family. Over this period Plato's reported job title advanced from "emp" (employee) in 1922, to "mgr" (manager) in 1923, "pres" (president) in 1925, and "pres and mgr" in 1929, this last designation suggesting perhaps less help if not less activity. Greater prestige did not necessarily mean more money.

The 1923 city directory revealed a name change that accounts for the promotion. Its classified business section listed two firms under "Rural

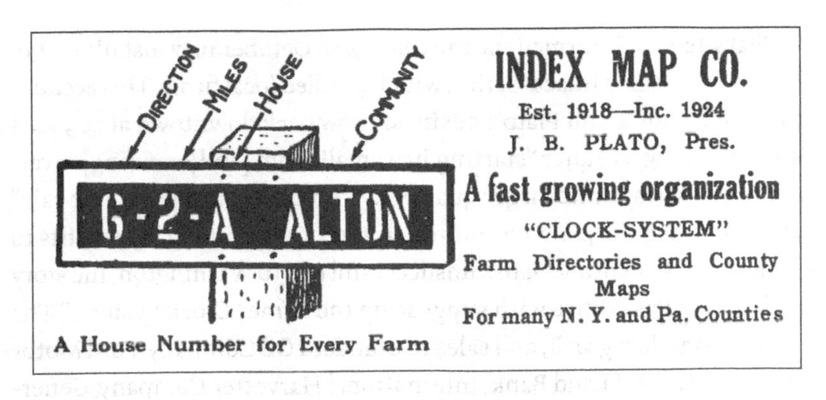

FIGURE 28. Quarter-page ad in the 1926 Ithaca city directory. *Ithaca (New York) Directory for the Year Beginning April 1926* (Schenectady, NY: H. A. Manning Co., 1926), 28.

Indexing," both employing Plato and both at the Calendar Clock location.[52] The newcomer was the Index Map Company: the result of a corporate restructuring in which Gannett was no longer president and the old board of local influencers was replaced by a new board of the company's three shareholders—Plato, his mother, and a local attorney, E. Morgan St. John.[53] Their one hundred shares of common stock were the only shares with voting privileges, and Plato held ninety-eight of them. He was president, he owned the critical patent, and he was now fully in charge. A key line in a quarter-page city directory ad ("Est. 1918—Inc. 1924") portrayed the restructuring as the culmination of a process started six years earlier (fig. 28). Eager to raise capital for his relabeled venture, Plato placed a subtly worded classified ad in the local paper: "INVESTMENT—A number of shrewd Ithaca business men are picking up a few shares of Index Map Co. stock."[54]

The *Ithaca Journal* largely ignored the new company. A one-paragraph story with an Albany dateline reported that the new firm had been authorized "to do an advertising business."[55] Although the story misspelled Plato's and his mother's last name as "Place," there is no reason to assume that Gannett harbored ill will. Nonetheless, the days of enthusiastic treatment in the local daily were over. The newspaper database I used found only three subsequent stories about the Index Map Company or its predecessor. One reported an auto accident involving a company employee, and another noted that Plato's plant was to host an "inspection trip" by the local chamber of commerce.[56]

That site-visit informed the third story, an October 1927 installment in the *Journal*'s "Know Ithaca" series, which profiled local firms. This account is rich in details about Plato's business, now back downtown at 209 East State Street (fig. 26) after "starting in a small room [and] mov[ing] several times . . . always into larger quarters."[57] After repeating the "lost sale" tale, describing the patented addressing scheme, praising its benefits to farmers, and lamenting Plato's unsuccessful trip to Washington, the story credited the Post Office with suggesting the name "Clock System." The business was doing well, and sales to Standard Oil Company, Ford Motor Company, Federal Land Bank, International Harvester Company, General Motors Company, Dodge Brothers "and hundreds of firms making or selling farm and household supplies" had fueled its "steady growth." The directories and maps helped a variety of businesses find their rural customers, and ads for national firms that listed local outlets helped farmers as well as advertisers. The Index Map Company's collaboration with local chambers of commerce "to bring cities and villages in closer contact with their surrounding territory" also made local advertisers aware of the rural directory's value.

Although the article revealed little about the firm's workforce, and who was doing what when the chamber visited, it reported that "during the busy season [the company] has about 20 employees." A concise overview of the workflow included transferring features from topographic maps with a pantograph and making blueprints for "men who go over every road and mark in the location of every house." Although the original business plan depended upon schoolchildren to collect data, "it was found best to have men hired to visit each farm." Cartographic field sheets were mailed back to Ithaca, where maps were drafted and farmers names sorted alphabetically for the typesetter. Printing plates were produced across the street at the Ithaca Engraving Company, profiled six weeks earlier in another "Know Ithaca" story.[58]

Because Clock System maps were typically published at the inch-to-the-mile scale of USGS topographic maps, tracing roads, boundaries, and other features directly onto transparent paper atop a light table might have been an efficient way to add this cartographic frame of reference. A light table is a large sheet of plate glass with a light source underneath: the source map is affixed to the glass with drafting tape, and the tracing paper is taped down on top. Light tables could be purchased from a draft-

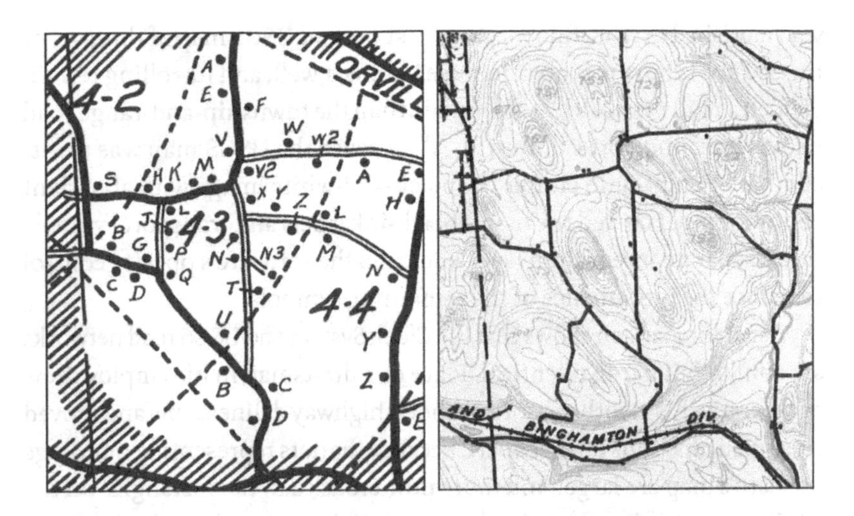

FIGURE 29. Comparison of a small part of the 1927 Clock System map for Onondaga County (left) with the corresponding portion of the USGS topographic map (right) on which its road network was based. The one-mile spacing of the circles (left) provides the maps' scale. Left excerpt from *"Clock System" Map of Onondaga County, New York*, series no. 16, ca. 1:72,000 (Ithaca, NY: Index Map Co., 1927). Courtesy Geography and Map Division, Library of Congress. Right excerpt from US Geological Survey, *Syracuse, New York* (quadrangle map), 1:62,500, 15-minute series, 1908.

ing supply firm or constructed by a handy carpenter—Plato no doubt had the necessary skill.

In the bygone era of pen-and-ink cartography it was customary to ink the final copy at an "up-size," from which a smaller, "down-size" negative was produced photographically with a copy camera. Photoreduction to a publication-size negative was an efficient strategy for minimizing minor irregularities in inking lines and labels on the final drawing.[59] For Clock System maps, these up-size ink drawings were probably based on the pantograph tracings used to produce the blueprints for the road-by-road canvass. A local engraving firm or commercial printer could have made both the blueprints and the photographic negatives.

To assess the geometric accuracy of Clock System maps, I extracted a small portion of the 1927 map for Onondaga County and used Photoshop to juxtapose it with a corresponding excerpt from the USGS's 15-minute quadrangle map of Syracuse, New York (fig. 29). Based on an 1893 field survey published in 1898, this was the most current USGS map of this part of the county. I chose this area because I know it well, and its rolling terrain yielded a road network less formulaic than the township-and-range road network around Fort Collins (fig. 3). Because the USGS map was printed in color and would clutter my black-and-white image with abundant contour lines, I moved the map's black-ink labels and transport symbols into a separate foreground layer and used the software's opacity control to reduce the prominence of the remaining symbols.

Visual comparison shows that the Clock System and USGS road networks are similar but hardly identical. Plato or a draftsman in his employ drew comparatively smoother, more rounded highway delineations and moved close road junctions farther apart. Because the dots representing buildings on Plato's map are larger and more numerous than the rectangles on the USGS map, shifting them farther apart and moving them farther back from the road are an appropriate exercise of cartographic license because the improved clarity outweighs any loss of locational precision. A typical user would count farmsteads onward from a particular road intersection and look for the occupant's address sign.

These metal gatepost signs were a valuable aid to motorists insofar as the Clock System had become well known, and almost all farmers had a car. Useful to be sure, but not trouble free. Although the firm had had "a great deal of difficulty . . . in getting a satisfactory paint for these numbers," it solved the problem by switching to an "auto lacquer," field tested on cars exposed to changing weather.[60]

Further evidence of Plato's success was a steady increase in the number of counties covered, as well as the release of revised editions, possibly every two years, to update address lists made obsolete "because many farms are rented now-a-days."[61] After mapping Tompkins County one or two towns at a time, he issued a directory for Cortland County, immediately to the east, followed shortly by directories for Tioga and Chemung Counties, to the south. The profile's author missed an opportunity to question Plato about the directory for Erie County, Pennsylvania, well west of the counties mapped.

To better understand the effects of proximity and distance, I plotted the counties covered by Plato's two firms on separate maps for easy comparison (fig. 30). For an efficient use of space, my base map includes only those New York counties covered by either the Clock System or the Compass System, implemented by another Ithaca firm in the late 1930s, after Plato had left

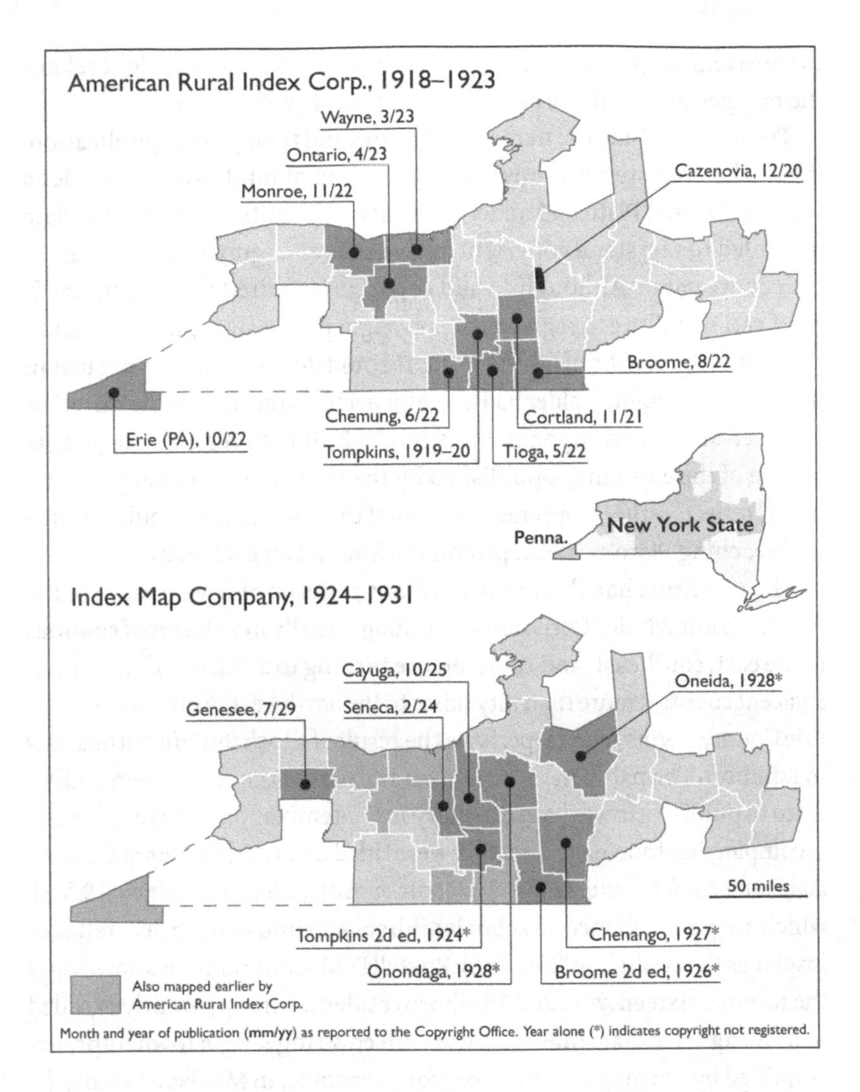

FIGURE 30. Geographic coverage of Clock System maps and directories, by county and publication date, shown separately for the American Rural Index Corporation (above) and the Index Map Company (below). Base map includes only New York counties covered by a Clock System or Compass System map; see inset. The dot in the center of Tompkins County is also the approximate location of Ithaca. Compiled by author from the *Catalog of Copyright Entries*; "Guide to Rural Indexes and Maps, 1920–1940, 1959" (Kroch collection number 3953), Cornell University, Rare and Manuscript Collections; and direct examination of maps in the Kroch collection. The Cornell finding aid is online at https://rmc.library.cornell.edu/EAD/htmldocs/RMM03953.html.

the area and his patent expired. A small inset map at middle right relates the mapped area to the broader boundary of New York State.

The maps include the names of counties and their year of publication. If Plato had registered a copyright, the month of publication precedes a two-digit year. If Plato had merely included a © notification and year date but failed to register a copyright, only the four-digit year accompanies the county name. Although a valid copyright required the customary © notification, failure to register the copyright promptly typically precluded a valid copyright unless the form, fee, and deposit copies were lost in transit and the rights holder had a receipt demonstrating that the materials had been submitted "upon publication."[62] Failure to register copyrights for five of the eight maps published by the Index Map Company (fig. 30, lower) reflects either sloppiness or a belief that an infringer could be sued for breaching Plato's patent, prominently noted on each map.

The two firms had distinctly different patterns of coverage, with the American Rural Index Corporation working initially in a cluster of counties to the east, southeast, and south before turning to a cluster of three nonadjacent counties more than fifty miles to the northwest. A clear anomaly is Erie County, Pennsylvania, perhaps the result of Clock System enthusiasts in Edinboro Township in the south-central part of the county persuading Plato to publish a town-level directory in September 1921. The map and its accompanying four-page directory were titled *Clock System Map of Edinboro Community, Erie County, Penna.: Territory Covered by Edinboro Vocational School*, which suggests reliance on schoolchildren—an unusually heavy reliance insofar as the credit line "Drawn by Russell D. McCommons" acknowledges the role of a sixteen-year-old Edinboro resident, who apparently excelled in drafting.[63] At that time Plato was still covering single townships, exemplified by the map and directory for Cazenovia, in Madison County. (A rural directory for the county as a whole was not published until the late 1930s.) By contrast, the Index Map Company added three comparatively close counties, three additional counties within a hundred miles, and two updated maps, for Broome and Tompkins Counties.

The maps reveal significant differences in county size and shape, which called for clever solutions to make efficient use of paper cut in rectangular sheets. For Genesee County, one of the smaller counties covered, the black-and-white Clock System map was at the center of a rectangular frame of advertisements printed in red and green (fig. 31).[64] By contrast, Cayuga

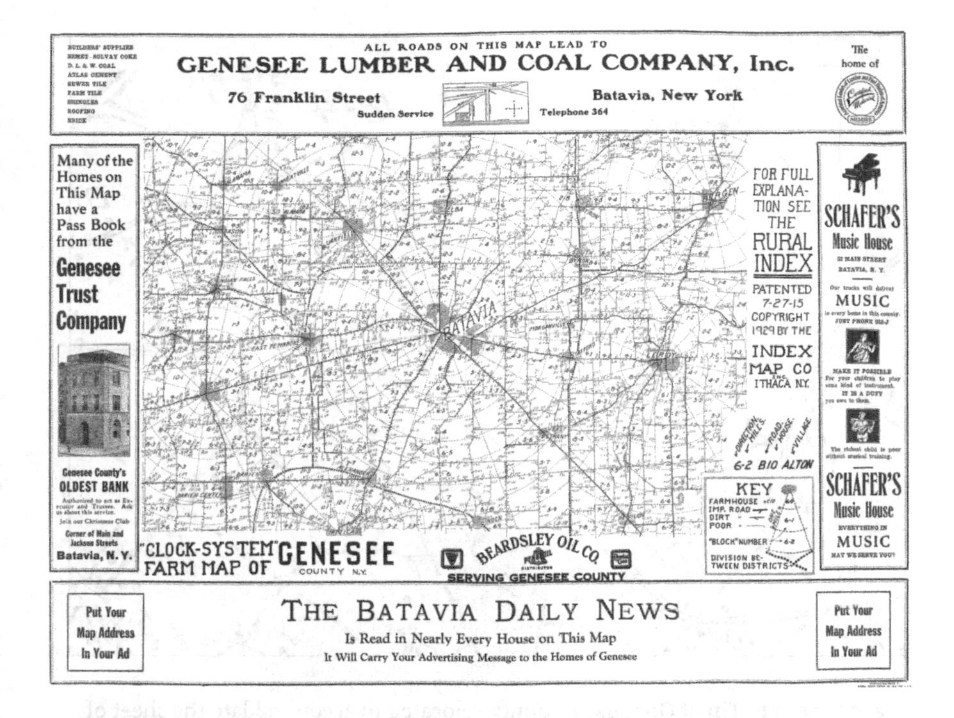

FIGURE 31. *"Clock System" Farm Map of Genesee County, New York* (1929) had a comparatively small, roughly rectangular shape that conveniently accommodated a frame of advertisements. 43 × 56 cm. *"Clock System" Farm Map of Genesee County, New York*, ca. 1:122,000 (Ithaca, NY: Index Map Co., 1927). Author's collection.

County, directly north of Tompkins County, has an elongated north-south shape, with a comparatively thin extension as it approaches Lake Ontario (fig. 30). To make more efficient use of the map sheet, Plato transferred the northernmost ten or so miles of the main map to the opposite side, and inserted "(OVER)" directly above the cut line, after having carefully avoided severing labels.[65] To make a minor bulge along the northern border of Onondaga County fit the paper, he removed a small section with eight farmsteads, placed it within a rectangle tucked in on the left (fig. 32), and added the note, "Continuation of roads shown at the right."[66] For Oneida County, to the northeast, he simply replaced most of the pointed northern extension with a smaller-scale map showing Long and Otter Lakes, a few roads and streams, and the village of Woodgate: a reasonable substitution because this part of the Adirondacks had no farms.[67]

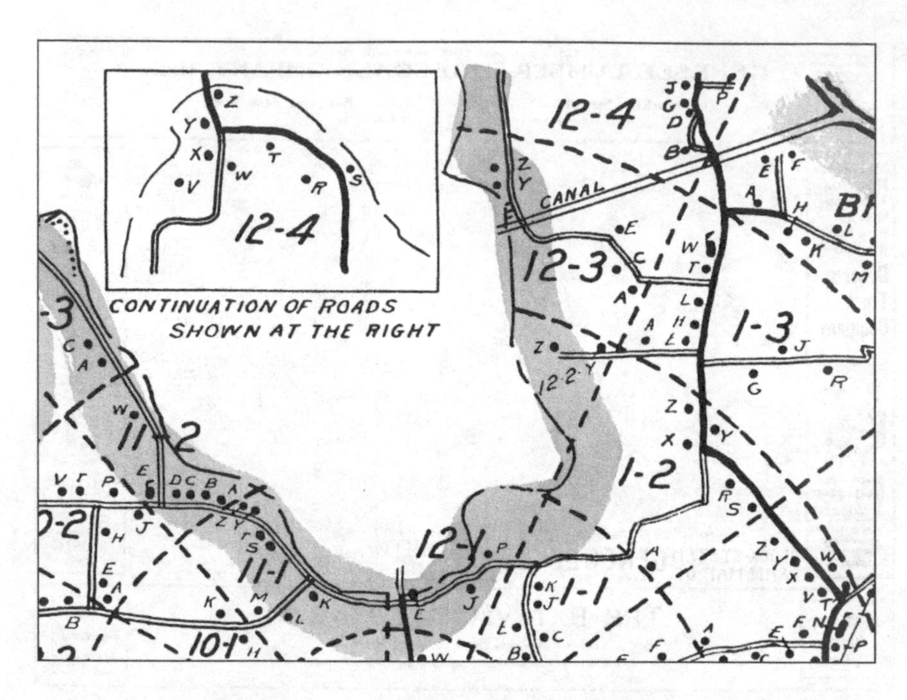

FIGURE 32. Tip of Onondaga County relocated to accommodate the sheet of paper. 8 × 11 cm. Excerpt from *"Clock System" Map of Onondaga County, New York*, series no. 16, ca. 1:72,000 (Ithaca, NY: Index Map Co., 1927). Courtesy Geography and Map Division, Library of Congress.

Sometime after publishing the Onondaga County map (fig. 32) in 1927, Plato introduced a new approach to labeling dots. His innovation, nicely exemplified on the Genesee County map released later that year, was a two-part dot label in which a letter representing a particular road was preceded by a sequential integer indicating relative position along that road. Figure 33, a full-size excerpt from the much-reduced whole-county version in figure 31, illustrates the concept. Note, for example, the paved road extending east-northeast from the village of Corfu: all farmsteads have dot labels that begin with B, starting with B5 on the western side of block 3-1 and extending beyond B57 on the eastern side of block 3-4 into blocks 8-6, 8-5, and 8-4, focused on Batavia, the county seat. Gaps in the sequence simplify renumbering if a new rural residence is added before the next revision.

In working up the graphic comparison in figure 29, I uncovered a more cartographically questionable strategy for shoehorning a map onto a paper

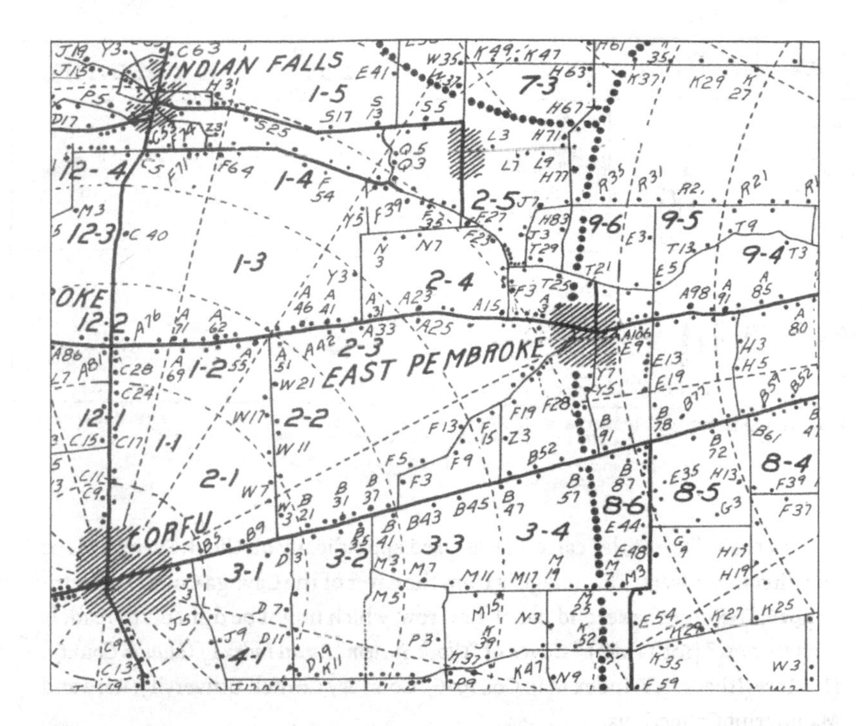

FIGURE 33. Excerpt from the Genesee County map illustrates the two-part dot labels consisting of a letter for the road and a sequential integer for the residence. Excerpt from *"Clock System" Farm Map of Genesee County, New York*, ca. 1:122,000 (Ithaca, NY: Index Map Co., 1927). Author's collection.

sheet. Although Plato's map for Onondaga County explicitly states in its key "SCALE—1 mile to the inch," a more accurate statement would be "1 mile to 7/8 inch." Whoever catalogued the map at the Library of Congress was sufficiently skeptical to make a few measurements, which revealed that the map's bar scale and the one-mile increment between adjacent circles were correct, but its verbal scale was significantly off. Instead of 1:63,360 (typically rounded to 1:62,500), the catalog entry, which requires a ratio scale, reads "ca. 1:72,000."[68]

Should this be called an error, a blunder, or merely an inconsistency? Take your pick. Although misstatements of map scale are unusual and hardly to be condoned, this anomaly is a useful illustration of the robustness of the graphic (bar) scale, in contrast to its verbal and numerical counterparts.[69] But on Plato's maps a bar scale is superfluous because the

FIGURE 34. "Lost sale" cartoon was used sporadically on Clock System maps and directory covers. This image is from the cover of the Cayuga County directory (1926). Text for the second panel, first row, which might be difficult to read, is "Where now? (10th time)." Cover of *"Clock System" Rural Index of Cayuga County, New York* (Ithaca, NY: Index Map Co., 1926). Courtesy Cornell University, Rare and Manuscript Collections.

consistent one-mile spacing of concentric circles around village centers provided a pervasive and fully reliable indication of distance.

Although clearly useful for promoting Clock System addresses, Plato's maps lacked the polish of products offered by most commercial mapping firms and government agencies of the time. This graphic informality is especially apparent in Plato's preference for hand lettering, rather than typeset paste-up labels or the unvarying characters produced with lettering templates or mechanical devices such as Wrico lettering stencils and pens, which guaranteed uniform, exactly spaced letters but were much more time-consuming than competent freehand labeling.[70]

However awkward aesthetically, Plato's lettering is fully legible. Moreover, its informal style aligns with the light-hearted, quintessentially practical attitude toward cartography reinforced by cartoons and folksy verse sprinkled about his publications. All carry a clear message. For example, his "Cow for Sale" cartoon (fig. 34), which he might have drawn himself, is a pointed parable illustrating the downside of the Post Office's vague directions. Similarly, his poem "The Tale of the 'Pig Tail Ad'" uses simple a-b-c-b rhyming for an engaging explanation of the Clock System's advantage over the RFD address, useful only because "It brings a letter to your house."[71] Its nine stanzas appear on many Clock System maps and directories and in newspaper ads:

Our farm house has a number now,
Just like the city folks,
It's based upon a wagon wheel,
With just a dozen spokes.

The village is the hub, of course,
With spaces all around.
Each space has got a number,
So each farmer can be found.

A farmer got it up one time,
By looking at a clock.
He tried to advertise one day,
To sell some growing stock.

To find the farm is just the same,
As telling time you know.
Each number's like a guide post.
To tell you where to go.

But the number on your mail route,
Don't tell a blooming thing.
It brings a letter to your house,
But that's all that it will bring.

You can't tell where a farm is at
By the "R. D." in an ad.
But the farmer had to use it,
'Twas the only way he had.

So a "pig tail" ad is one that ends,
In the letters "R. F. D."

It's called a "pig tail" ad, because,
It points nowhere, you see.

"It's the old Brown farm, on the second road
From the hill by Jasper Jones,
You'll know the place by the big red barn,
Where the road is full of stones."

But now we've got our number,
To go upon our gate.
We've got our "Index" and a Map
And now we're up to date.

—J. B. Plato[72]

Plato had tweaked his business model by charging farmers twenty-five cents for the Index and fifty cents for the map but offering "a Special Price to Farmers if Purchased from Rural School Children at time of putting up FREE number plates."[73] Free signs promoted a widely visible presence that made the Clock System a reliably ubiquitous navigation aid as well as a subtle but unavoidable advertisement likely to enhance sales. And who can resist an entrepreneurial youngster performing a public service? To further shore up his revenue stream, he expanded the product line with a new publication pitched to "Merchants and Manufacturers." The earliest might have been the *1926–27 Special Classified and Graded List of All Farms of Broome County for Merchants and Manufacturers*, which identified farmers as owners or tenants; categorized farm types as general, dairy, poultry, "both dairy and poultry," fruit, and truck; and graded farms by size and appearance as:

F for Fine places.
A for Above average in appearance.
R for Regular farms.
M for Smaller places.

F A R M—get it? With no obvious pejoratives apparent, the "Classified and Graded Directory" was pitched to sellers "for circularizing, for follow-up work, for demonstrations, for credits, for deliveries, for billing, for collections, etc."[74]

Although Plato told the *Journal* he employed twenty people during the busy season, my hunch is that most were part-time or temporary hires.[75] His March 1922 "Help Wanted—Female" ad in the *Ithaca Journal* said as much by expressing a need for "GIRLS—To do typewriting. Stenography not necessary."[76] The only employees I could identify by name were the aforementioned student drafter in Edinboro, Pennsylvania; Loran Baker, a local freelance sales representative who sold advertising and helped compile the first Broome County map in early 1922;[77] and Frank L. Tyler, who was corporate secretary of the Index Map Company in 1929 and ran his own real estate business.[78]

A small intermittent workforce was consistent with Plato's modest workspace and storage needs. According to the 1927 city directory, the Index Map Company shared room 304 on the third floor at 209 East State Street with a sign studio and a detective agency.[79] Other tenants on the floor were an elocution teacher and Cornell Annuals, which published the school yearbook. Plato's formula for keeping space, labor, and operating costs low was consistent with his low-budget accommodations in rooming houses—an easy formula for quietly succumbing to an economic tsunami.

Agriculture and commercial cartography suffered massive retrenchment during the Great Depression following the Stock Market Crash of 1929. Farms were burdened with debt, and overproduction had undermined profitability. Diminished consumer confidence forced map publisher George Clason out of business in 1931, and the Index Map Company, so closely tied to agriculture, was doubly vulnerable. The firm registered its last copyright in 1929 and released no new maps or directories in 1930.

Although Plato's few shareholders were no doubt disappointed, his business was apparently too small for a dramatic closure or newsworthy bankruptcy. I found no evidence of an auction or sale of equipment and inventory, or an attempt to sell the firm. Shortly after Plato's mother died at age eighty-nine in a local hospital in early March 1931, he closed the business, vacated his lodgings, and left town.[80] The city directory for 1931 reported John and Helen Plato as roomers at 201 Center Street,[81] but the 1932 directory, which had dropped mother and son, still listed the Index Map Company at 209 East State Street, now sharing room 304 with a floor refinisher.[82] In the next year's directory room 304 was completely vacant.[83]

6 OHIO

Plato did not abandon farm mapping when he closed his business in Ithaca. In much the same way that he transitioned into and out of farming at Semper, Colorado, during the 1910s while moving among multiple residences, he initiated and then spun down a new endeavor in Ohio before moving on to federal employment in Washington, DC. Reemergence of his pattern of overlapping residences was revealed in the *Findlay* (Ohio) *Morning Republican* for July 16, 1930, in a six-paragraph story under the headline "Farms Are Given Street Numbers; Hancock County Directory to Describe Location of Farms by Number."[1] Plato still had housekeeping rooms at 201 Center Street in Ithaca, where his mother Helen would live for seven more months. He was still renting office space in downtown Ithaca, and yet, here he was, nearly 400 miles to the west-southwest, mapping farms, soliciting ads, and publishing rural directories.

How Plato wound up in Findlay is a mystery. Hancock was a farming county, to be sure, but no more distinctive than its neighbors.[2] Plato had won the all-important support of the Findlay Chamber of Commerce, and by early September a large map explaining his system was in the window of the C. W. Patterson and Son department store in downtown Findlay.[3] Though press accounts never mentioned the "Clock System" by name, they described the numbering scheme based on direction and distance from locally significant villages: the same principles Plato had applied for over a decade in New York State. Moreover, the map was to be published by the Index Map Company, his old firm in Ithaca, and the Hancock County Farm Bureau would distribute free "number plates" with Clock-type addresses.

Two months later, on September 12, an ad in the local newspaper noted that the "new rural index and county farm map" was "being made ready for printing" and listed the fifty-eight advertisers "among the first group of firms co-operating."[4] Interested parties were invited to write "Mr. J. B. Plato, Box 245, Findlay," for further details. Post Office box numbers gave Plato flexibility for his business and himself. Further confirmation that he had moved to Ohio was an instruction in the assessor's office in Colorado routing the 1931 tax bill for his 0.4-acre lot in Semper to "Johnstown, Ohio, Box 25."[5] He used the Johnstown address until 1936, when his tax warrant was sent to a Washington, DC, RFD address.

In late January a news story headlined "Farm Directory Completed Here" confirmed that printing was imminent.[6] According to the *Morning Republican*, the directory used a multi-factor classification scheme that coded a farm C for cattle or feedlot, D for dairy, F for fruit, G for general, H for hogs, P for Poultry, S for Sheep, and T for truck. In addition, an R indicated a non-farm dwelling such as a tenant's house or a "home in the country that is not a farm." In addition, properties were assigned a quality rating of X for "excellent or above average," Y for "average," and Z for "other"—a polite way of saying worse than average. A capital X, Y, or Z indicated "owned in the family," whereas a lowercase *x*, *y*, or *z* signified a tenant. Several years later, when Plato sought work at the Census Bureau, his practical experience in farm classification was his primary qualification. Because of detailed news stories like this, I know the Hancock County rural guide exists, even though I could not find a copy of the map or the directory in any library, archives, or map collection. Plato never registered any of his Ohio rural directories for copyright.

As facts began accumulating, I recognized that a map was essential for organizing information about where Plato lived and published maps in Ohio. An inset in the lower left corner of figure 35 describes his area of activity in north-central Ohio. The map shows all of the places connected with Plato between 1930 and 1933. I added Columbus, the state capital, to complete the geographic frame of reference, and drew in the boundaries of the four counties that Plato is known to have mapped—there could be one or two others.[7] To promote visual ambience, I included a 1931 Ohio highway map in the background layer, suitably faded to avoid graphic conflict. Hancock County, and its principal city, Findlay, are at the upper left, and Johnstown is at the bottom center, within Licking County and more than 15 miles west-northwest of the county seat, Newark.

Plato's PO box address in Johnstown signaled the selection of Licking County for his next cartographic project. Nine months after release of

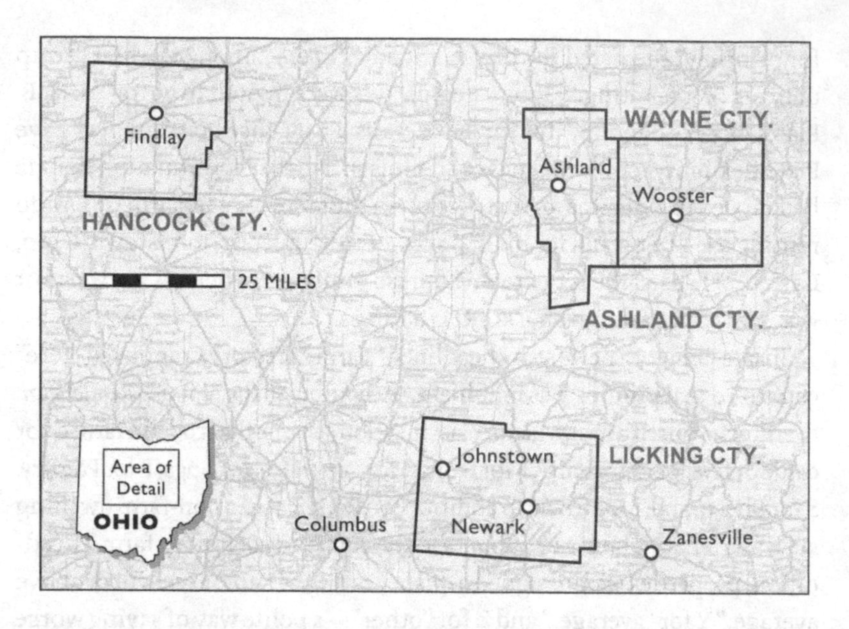

FIGURE 35. John Byron Plato's residential and work locations in Ohio, 1930–1933. Map in the background is State of Ohio, Department of Highways, *Map of Ohio Showing State Highway System*, 1931, original scale approximately 1:750,000. Foreground compiled by the author.

the Hancock County directory, the *Newark Advocate* for October 14, 1931, reported that mapping was well underway. A "field crew" had gone from "house to house over the 1,400 miles of Licking County roads" to locate each farmhouse, acquire the farmer's name, and record the type of farm and whether it was on the electric power grid.[8] Once drafting was complete, the directory would be distributed free, by mail, to all farmers in the county. Accompanying the directory were two coupons: one for a free copy of the "new Farm Index Map," which could be picked up at a local bank or store, and another for a metal "number plate" with the farm's rural address.

The new distribution plan, with maps and directories bound and distributed separately, reflected the establishment of a new firm, Farm Index Service, Inc., in October 1931, in Wooster, the county seat of Wayne County (upper right in fig. 35).[9] An expansive charter described a business authorized to "make, manufacture, buy and sell and generally deal in maps, mailing lists, directories, sign boards, bulletin boards, sign posts and all articles used or useful in connection therewith . . . and to do a general advertising business." To raise capital, the firm could issue one hundred shares of preferred stock, each with a par value of $100, and one hundred fifty shares of "no par value" common stock. Only the holders of common shares were allowed to vote. That the corporation would begin business with only "Five Hundred and no/100 Dollars" underscored the importance of selling shares and borrowing. I found no record of how many shares were sold or how much he paid out in dividends.

The articles of incorporation indicated that Plato had recruited two local partners: Edgar R. Miller, a high-school coach and principal in nearby Dalton, and Lyman R. Critchfield Jr., a prominent local attorney, son of a former judge, and grandson of a former state attorney general.[10] I found no indication that either was actively involved in the business.

Wayne County would have its own map and directory the following year, as would Ashland County, immediately to the west. I know of the Wayne and Ashland directories because copies were acquired by libraries that reported their holdings to the National Union Catalog and the Ohio College Library Center (OCLC), which became the basis for the electronic database WorldCat.[11]

The Ashland directory and map were comparatively easy to locate because they had been acquired by the Public Library of Cincinnati and Hamilton County, which cheerfully scanned the copies in its collection. Oddly, neither WorldCat nor the National Union Catalog listed a holding library for the Wayne County directory, but an email query located a copy in the county library's genealogy department. An intact coupon (fig. 36) suggested the library never bothered to claim its free copy of the map. Email queries to county libraries and historical societies failed to locate the presumably similar publications for Hancock and Licking Counties, which I know of only because Plato talked to the press.

The coupon was embedded in the forty-eight-page 14 × 22 cm directory mailed free to rural residents of Wayne County. With at least two ads on every page, the directory was a typical saddle-stitch booklet. At the upper right of the front cover a "U.S. Postage PAID" bulk mailing permit referenced "Permit No. 15" on file at "Johnstown, O." Below the title Plato's oft-used "lost sale" cartoon illustrated "How the Rural Index took the 'Far' out of 'Farming'" (fig. 34). The coupon occupied back-to-back portions of the lower half of pages 13 and 14. The front side (fig. 36) listed over fifty busi-

Map Coupon

For FREE FARM INDEX MAPS, put your name on other side and take this coupon to one of the following. They are doing this to show their appreciation of your business.

APPLE CREEK— Wayne Farmers' Sup. Co. Stauffer Poultry Farm

CRESTON—
Gensemer Bros., Inc. Tyler Chevrolet Sales Plank Elevator, Feeds
Mast Hatchery, Sterling Sellers & Ober, Hardware N. J. Elliott, Madisonburg

DALTON—
Camp's Service Station First National Bank Haverstock & Davis, Ford
Orr's Pharmacy C. R. McDowell, Hdw. Meyers Restaurant

DOYLESTOWN—Doylestown Business Men's Association, All Members.

FREDERICKSBURG—
J. C. Murray, Funeral Ser. H. A. Stucker, Chevrolets J. C. Weaver, Gen. Store
E. E. Crosby Goody. Tires Hoffer Hardware Armstrong's Sales Barn

KIDRON—
Sprungers Auction Sales E. P. Gerber, Lumber-Hdw J. P. Sommer "Kidron" F.

MT. EATON—
Schaffter Bros., Imple. The Bank of Mt. Eaton E. E. Tschantz, Hdw. Fee.
Scheuftler & Tracy, Groc. Younkers Garage. AAA H. W. Desvolgnes, Furni.

ORRVILLE—People's Savings & Loan Company, Orrville Branch.

RITTMAN—
Rittman Hardware Co. Rittman Savings Bank Brenneman's Grocery
Rittman Auto Laundry B. & W. Oil Co., Com. Ser.

SHREVE—
Clarence Turner, Ins. George Wise, Grocer
Fought's Bakery Fought, Chevrolets

SMITHVILLE—
Scheller Bros., Livestock Farmers & Merchants Bk. E. E. Zimmerman, Grocer
J. J. Buchwalter, Mortl. Komfort Corner, Ser. Sta. M. & R. Garage, Fords

WEST SALEM—
W. Salem Equity, Feeds W. Salem News, Newspn. B. W. Mowery, Fun. Dir.
Zehner & Martin, Fords Farmers State Bk. of W.S.

WOOSTER— Citizens National Bank

FIGURE 36. Map coupon in the Wayne County directory listed multiple businesses at which the bearer could obtain a free county map. *Rural Index and Buying Guide for Wayne County, Ohio, for 1931 and 1932* (Johnstown, OH: Farm Index Service, Inc., 1931), 13.

nesses in thirteen communities at which the coupon could be exchanged for a county map. On the back a simple, generic request—"Gentlemen: I would very much appreciate having a copy of the new Farm Index Map of this County. Thank you."—was positioned above a dotted line for the resident's name. Additional copies of the map could be purchased for fifty cents. A slightly smaller coupon immediately above could be used to order a free 6 × 9-inch number plate with the farm's "New Address" printed in "non-rusting metal with a genuine Duco finish to place on the road in front of your home."[12]

How might a county's boundary affect the appearance of its map? Although address plates and map symbolization were no doubt similar for both counties, Ashland County's pronounced north–south elongation as well as its erratically angular border offered more abundant opportunities for small ads and other marginalia than Wayne County's simple rectangular boundary (fig. 35), which better matched the typical shape of a printed

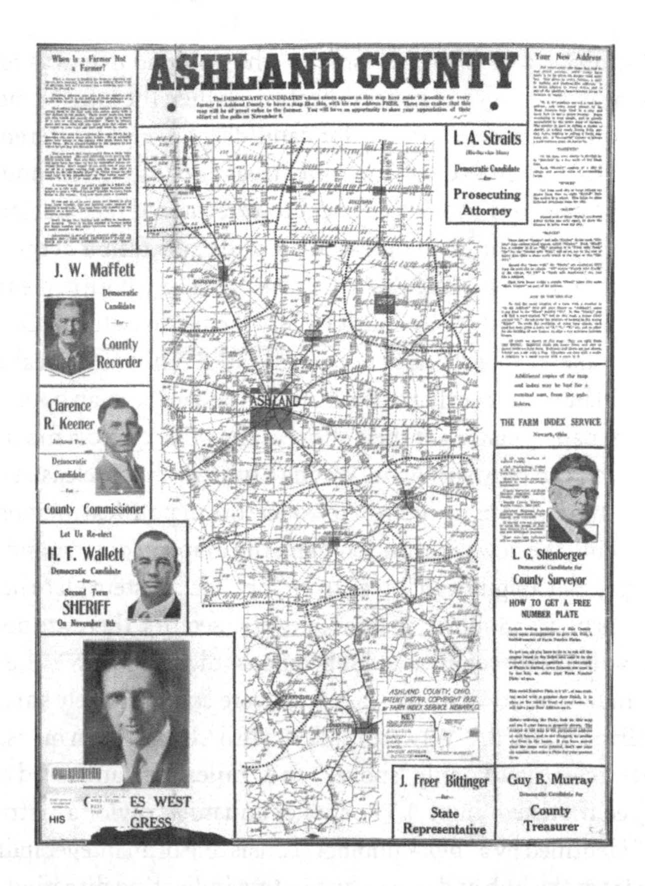

FIGURE 37. The Ashland County map doubled as an advertising vehicle. Reduced substantially from a published size of approximately 54 × 75 cm. Map related to *Rural Index and Buying Guide for Ashland County, Ohio, for 1932 and 1933* (Johnstown, OH: Farm Index Service, Inc., 1932). Map is from the online digital collections of the Public Library of Cincinnati and Hamilton County, Genealogy and Local History Department. The file name for the item is "Rural index and buying guide for Ashland County 977.129 R948.pdf." The URL for the directory is https://digital.cincinnatilibrary.org/digital/collection/p16998coll15/id/419488, and the URL for the map is https://digital.cincinnatilibrary.org/digital/collection/p16998coll9/id/7993/rec/2.

sheet. Timing became relevant insofar as the Ashland map was released between 1932's primary and general elections, when the local Democratic organization bought separate ads for candidates in eight different local races, from US Congress down to county treasurer (fig. 37). Although the space might have been sold piecemeal to tractor and feed dealers, text below the map title outlined the rationale and expectation of the party's combined ad buy: "The DEMOCRATIC CANDIDATES whose names appear on this map have made it possible for every farmer in Ashland County to have a map like this, with his new address FREE. These men realize that the map will be of great value to the farmer. You will have an opportunity to show your appreciation of their effort at the polls on November 8."

Although I can only speculate, Wayne County's map, released in 1931, apparently held little or no interest to collectors of political memorabilia.

Reinforcing the new distribution plan, the map's geometry and terminology represent a significant break from the Clock System Plato had used in Ithaca. As figure 38 shows, instead of twelve sectors, there are now only eight: the well-known compass and subcompass divisions (N, NE, E, SE, S, . . . And if a trade area, now called a "district," were comparatively small, only the four divisions (N, E, S, W) were used. As with Clock System maps, sector lines and circles (now simply "spokes" and "miles") are anchored at trade centers (rechristened "hubs"). The spokes and miles divide a district into "blocks," identified by a "block number" consisting of an integer indicating distance from the hub and one or two letters indicating direction. Within the southeast sector, for instance, the first block, within the one-mile circle, was 1SE, and subsequent blocks were 2SE, 3SE, and so forth. But if 1SE were wholly within the built-up or city area, block numbering would start with 2SE, as in the lower part of figure 38.

A more prominent modification of the design is the use of a second printing ink, red, to better differentiate block numbers and the grid of miles and spokes. Cartographic license allowed me a reliable description in figure 38 of the red overprint's visual impact. Had I merely converted the color map to a graytone graphic, its red lines and letters would have faded into a mediocre medium gray. The use of Photoshop helped to highlight the map's red symbols. After moving the reds and the blacks into separate layers, I assigned solid black to the smaller number of red features and medium gray to the more numerous black lines and labels. This substitution reduced significantly the prominent black district la-

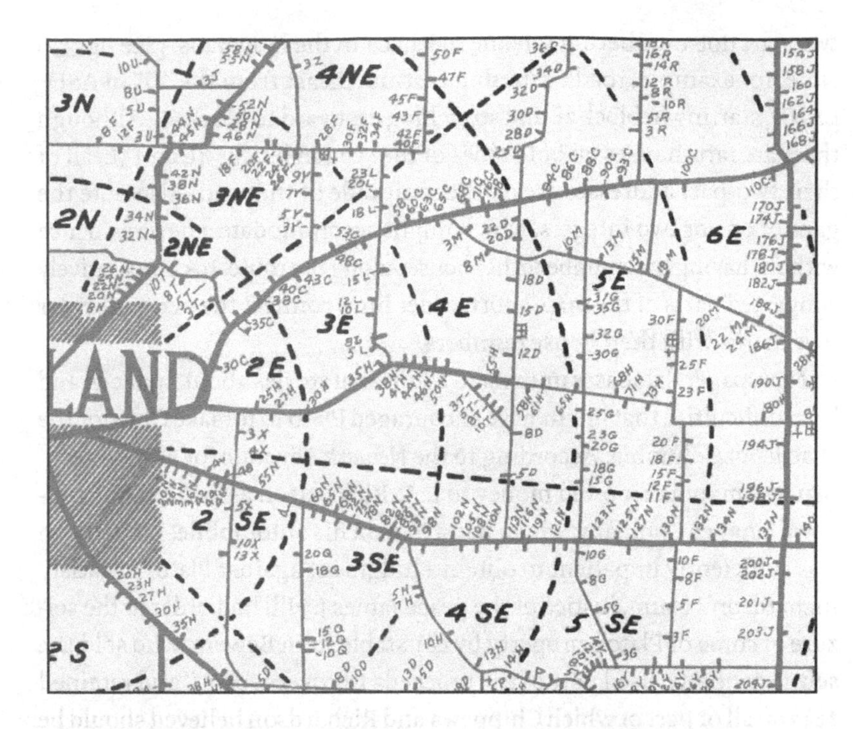

FIGURE 38. Enlarged excerpt from the Ashland map uses solid black for symbols originally printed in red and medium gray for everything else. Enlarged excerpt extracted from Figure 37.

bels and district boundaries at the top of the original graphic hierarchy. Because the spokes' red dashes still looked recessively thin, I thickened them slightly with an image filter.

Plato made further changes to content and symbolization. Note that thick black lines representing "improved roads" contrast markedly with thinner, more common lines representing "dirt or gravel roads." He also replaced the tiny dots representing farmsteads with short strokes or tick-marks perpendicular to the road to indicate the approximate location of each rural residence or its driveway, as well as its position (left or right) along the road.

Roads had an explicit role in the numbering of rural residences. Each road or string of residences is represented by a letter, which "runs" through multiple blocks and across district lines, while each house has an integer indicating its relative position along the route. Plato had introduced this two-part dot-numbering scheme in Ithaca in the late 1920s. (See fig. 33.) Note, for example, road C running east-northeast from the "D" in ASHLAND, starting in block 2E and stretching eastward to block 6E. Although there are farmhouses on both sides of the road in blocks 4E and 5E, all of their two-part addresses are on the same side of the road. Also note the gaps of one or two integers, which might accommodate a new residence without having to renumber other houses along the route. In comparatively congested parts of the map, short leader lines connect the perpendicular tick-marks with their house numbers.

Newspaper databases uncovered additional stories about financial and legal difficulties that might have encouraged Plato to forsake Ohio for the District of Columbia. According to the *Newark Advocate* for November 9, 1931, the mapmaker owed money to L. L. Richardson as well as the Chippewa Finance Company, which had a lien on his automobile. Richardson was sufficiently impatient to obtain a judgment against Plato. To satisfy Richardson's claim, justice of the peace James J. Hill had ordered the seizure of some of Plato's property by constable John Bowers, who sold the seized property, used part of the proceeds to cover "costs," and retained $247.61, all or part of which Chippewa and Richardson believed should be theirs.[13] Richardson argued that because he had obtained the judgment, all the money should be his.[14]

Plato claimed that the seizure was illegal. The warrant had been executed at the office of Farm Index Service, Inc., in Johnstown, and some of the items belonged to the corporation, not to him personally. Moreover, the seized property included "tools and implements which he used in his work," and should have been exempt from seizure.[15] He also argued that the "Ford sedan was necessary for carrying on his work."[16] Oddly, the court docket identified constable Bowers as the plaintiff, and "Chippewa Finance and others" as the defendants. As one of these "others," Plato was no doubt pleased when the judge ruled for the defendants.[17] Because he had been representing himself, the news report listed him among the defendants' attorneys, along with Critchfield, McSweeney, and Critchfield, the firm of one of his two co-owners, which was representing Chippewa Finance. Small world, indeed.

Leery of further seizures, Plato sought bankruptcy protection, which was granted on January 13, 1932. A legal notice in the *Newark Advocate* announced a meeting of creditors scheduled for February 15, in Zanesville, approximately twenty-five miles east southeast of Newark (fig. 35).[18]

That November Plato suffered a further indignity. Because Farm Index Service, Inc. had apparently failed to file an annual statement of earnings and pay whatever tax might be due, Ohio's Secretary of State canceled its articles of incorporation.[19]

In March 1933, still aggrieved by what he considered an illegal seizure and sale, and perhaps emboldened by his victory in the Bowers case, Plato filed a lawsuit against Hill.[20] Committed to keeping expenses low, he again acted as his own attorney. (He might have felt empowered because his father and grandfather were lawyers.) Hill argued that because Plato was still covered by his bankruptcy filing the previous year, his claim against the town justice, whatever its merits, was "an asset of his estate in bankruptcy," which gave the right of collection to his bankruptcy trustee. That is, if a lawsuit were to be filed, the only valid plaintiff was Plato's trustee.[21] Plato responded by arguing that "the subject matter of [his] petition is not and never was with the jurisdiction of the defendant as justice of the peace, his agent or the trustee in bankruptcy of [Plato's] estate."[22] In May the court ruled for Hill, whose action "was a judicial one [for which] he could not be held personally [liable]."[23]

Plato could not let the matter rest. In late October 1933, the case was back on the docket, this time for a jury trial. As before, he was representing himself. Because this was also the last time the *Newark Advocate* reported on his litigation or mapmaking, Hill might have settled the case before Plato left Ohio—or mercifully dropped the matter.

7 WASHINGTON, DC

After four years of producing maps and directories for several Ohio counties, Plato went to Washington for five years of intermittent work at the Census Bureau and the Agricultural Adjustment Administration (AAA), a division of the US Department of Agriculture (USDA). This latter phase included a brief and emblematic attempt in 1937 to initiate a Clock System map for a northern Louisiana parish.

Although the dates of Plato's departure from Ohio and his arrival in Washington are unknown, his residence in the national capital by early spring 1934 is well established by an article in the *Washington Evening Star*.[1] Proud of his service in the Spanish-American War, Plato sought affiliation with a local veterans group, known as a "camp," and on Thursday evening, March 22, "John B. Plato, First Colorado Infantry" and one other veteran were "mustered in" by the Lieut. Richard J. Harden Camp. The evening's program included talks by seven senior members.

Plato would not show up in a local city directory until the following year, when *Boyd's District of Columbia Directory* for 1935 had two listings for the same John Plato: a "Plato J B," residing in Forestville, Maryland, was employed as a "spl agt" [special agent] at the Census Bureau, and a "Plato John B," living in Benning, a DC neighborhood, was employed as an "asst agrl economist" at the AAA.[2] Created as part of President Franklin Roosevelt's New Deal, the AAA sought to offset plummeting farm income by eliminating surpluses, at first with a tax on processors and later by paying farmers to take land out of production.[3] Too productive for their own good, farmers had been harvesting ever more acres of cash crops, such as corn, and the inflated supply drove the price per bushel well below the level needed to cover production costs. Although Plato worked for the AAA through 1937, the 1936 city directory included only his affiliation with the Census Bureau and the Forestville address.[4]

Federal Civil Service records obtained from the National Archives reveal the shortcomings of city directories and newspaper databases. On November 27, 1934, in completing a job application for the AAA, Plato responded "13 months" to the question, "If now residing in the District of Columbia, how long have you been so residing?"[5] His answer placed him in the area in October 1933, no doubt to explore expanding Civil Service opportunities following Roosevelt's inauguration in early March 1933. Agricultural economist Charles Galpin, an early supporter of the Clock System who left the University of Wisconsin in 1919 to direct the USDA's rural life program, would not retire for another year.[6] Networking apparently paid off because on December 21, 1933, Plato was a federal employee, working at the USDA as an administrative assistant with an annual salary of $2,600, and the following summer he was also a Special Agent at the Census Bureau.[7]

Although the job title Special Agent sounds impressive, it was largely a bureaucratic ploy for acquiring expert advice at little or no cost. In Plato's case, the Bureau gladly authorized a salary of "$1.00 per annum"—the other option was "without compensation"—and provided office space for "the father of the farm identification scheme," who was assigned to the Agriculture Division to advise on the Bureau's "farm identification proposal."[8] Plato's role was probably related to the *Census of Agriculture*, a statistical compilation issued every five years. His appointment ran from August 1, 1934, through November 10, 1936, when "work completed" was reported as the "cause" of his "termination of service."[9]

Plato's appointment, which included an oath of office, was based on qualifications summarized in a two-page "Personal History Statement."[10] Under specific qualifications, he wrote, "Special Experience in Farm Identification in connection with [the] New York College of Agr[iculture] at Cornell," and under principal non-federal employment, he reported having served for fifteen years as president of the Index Map Co. Under education, which merely asked the applicant to circle the number of years of "Common school," high school, or college, he circled "1" for the latter and wrote in "Cornell." His eleven-week Winter Course was now a year of college.

Well aware of the value of the Cornell brand, Plato occasionally embellished his connection with the university. On a July 1935 application to the USDA's Resettlement Administration, he invoked the time intervals

1895–96 and 1918–32 to describe his education as "Special work at Cornell University."[11] A month earlier, on an application to the same agency, he filled in the "any special qualifications" section with "Lecturer at Cornell on Farm Identification,"[12] and a year earlier, in a statement for the AAA, he had pinned down the lecturer position to the year 1925.[13] (Although Plato might have been an occasional guest lecturer, I found no record of his having held a faculty position.) Not averse to name-dropping, he had once described his "special work at Cornell" as study "under Prof. G[eorge] S. Warren, authority on farm identification."[14] A prominent agricultural economist, Warren was one of Roosevelt's key advisors on agricultural issues.

None of these embellishments was as egregious as the "employment record" filled out in May 1934 by a USDA administrative officer who wrote, "[he] has been employed for the past 15 years at Cornell University working on a project in connection with the New York State Department of Agriculture (1918–33)."[15] Whoever had typed up the form had apparently reported whatever Plato told him.

Plato held a government job more or less continually from late 1933 through October 1937, mostly as a series of one- or three-month appointments punctuated by formal letters of termination quickly contradicted by a reappointment, sometimes labeled urgent. Although his job title changed from Assistant Agricultural Economist in 1933, 1934, and 1935 to the more ambiguous Junior Administrative Assistant in 1936 and 1937, his consistent annual salary of $2,600 was well above the 1935 averages for federal civilian workers ($1759/yr) and public school teachers ($1,293/yr).[16] Average managerial and professional salaries were no doubt higher, but he clearly enjoyed a comfortable income, albeit well below the $6,000 "highest salary" he ascribed to his employment as president of a map publishing firm.[17]

A significant disruption occurred in May 1934, when the AAA needed clerks more than it needed assistant agricultural economists. Plato's salary dropped to $1,800, commensurate with the position of Clerk, and a month later to $1,260, for the lesser job title Under Clerk. A windfall of data that might yield a valuable analysis had apparently created an urgent need to reformat railway freight records on manufactured commodities shipped into areas affected by AAA activities. According to the director of the agency's Division of Information, "The utmost haste is necessary in performing this work because unless the data are turned over quickly to the Contract Records Section that section will be unable to spare the use of the machines for mechanical tabulation."[18] (Although *computer* still referred to a human who relied on an adding machine or logarithms, IBM machines were adept at counting punched cards.) To expedite hiring, the AAA had created temporary jobs outside Civil Service, and conveniently reassigned Plato to one of these emergency positions, in which he would "under immediate supervision . . . receive raw statistical data . . . code and assemble these data for mechanical tabulation and analysis, and . . . perform related work as assigned."[19]

How long the redeployment lasted is not clear. Civil Service records in the National Archives include five short-term appointments: the first on May 2, 1934, and the last on July 31, 1935. According to an investigator's report, "The applicant was appointed by the AAA as a clerk at a salary of $1,260 on June 15, 1934. He was later promoted to assistant agricultural economist at a salary of $2,600, and on May 31, 1935, he was transferred to [the] Resettlement Administration. He is now serving as assistant economist under Dr. L. C. Gray at a salary of $2,600."[20] The report described Plato as "a very high type man, capable, [and] efficient, [whose] present services are very satisfactory."

By July 1935 Plato had a job description consistent with his experience:

> Under general direction of the Chief of the Land Use Planning Section, with considerable latitude for the exercise of individual initiative and judgment, to continue the conduct of research in the application of a farm identification system upon which he has been working exclusively for some years; to develop and improve the farm identification methods thus far worked out; to prepare maps and directories embodying the application of the methods, and to study their usefulness to various governmental agencies; and to plan and propose projects for the extension of farm identification methods to different parts of the country.[21]

Plato's unique interest in farm identification obviated a typical positive personnel evaluation. According to his supervisor, Dr. Gray, because he was "the first person to do any farm identification work and the type of identification upon which he has been working is his own idea, we . . . find it impossible to form a comparison upon which to base a rating of his work." Gray added, "His idea of farm identification is good, his development of the idea has been satisfactory to us and has been utilized to some extent by the Census Bureau." But because Plato's "interest and training lies almost wholly along the lines of farm identification"—Gray's wording suggests excessive specialization—officials in the Land Utilization Division were reluctant to offer him another position.

In November 1936 the AAA posted Plato to its field office in Athens, Georgia, as a Junior Administrative Assistant at his customary salary of $2,600, with a new job description that downplayed farm identification:

> Under general supervision, Mr. Plato will prepare, with the assistance of the Directors of Extension, County Agents, Assistants in Agricultural Conservation, County and Community Committeemen, maps of the counties, locating thereon the farms of such counties and assigning permanent numbers and identification to each farm; develop and establish a standard procedure for the entire region with respect to the permanent identification of farms therein located; thereafter instruct and supervise County Agents, Assistants in Agricultural Conservation, County and Community Committeemen in the preparation of maps and the identification thereon of the farms throughout the States of the Southern Division.[22]

He would interact with the county committees and local officials involved in the AAA's monitoring and mapping of farms: an essential strategy for making certain that farmers receiving AAA payments honored their commitment to take land out of production. Oddly, his job description said nothing about aerial photography, which county committees used to check on their neighbors' planting and harvesting of crops.[23] The work apparently focused on making the basic maps used to locate and account for all farms in a county, and on showing AAA collaborators throughout the South how to make their own maps.

To Plato, the job description was an invitation to revive the Clock System, or at least the modified eight-sector version used in Ohio. Although neither the National Archives nor the USDA History Office had a record of his activities—we know when and for what he was appointed and when he was terminated, but very little in between—three newspaper stories document a bold and carefully orchestrated effort to apply familiar cartographic principles in Lincoln Parish, a rural county in northern Louisiana,

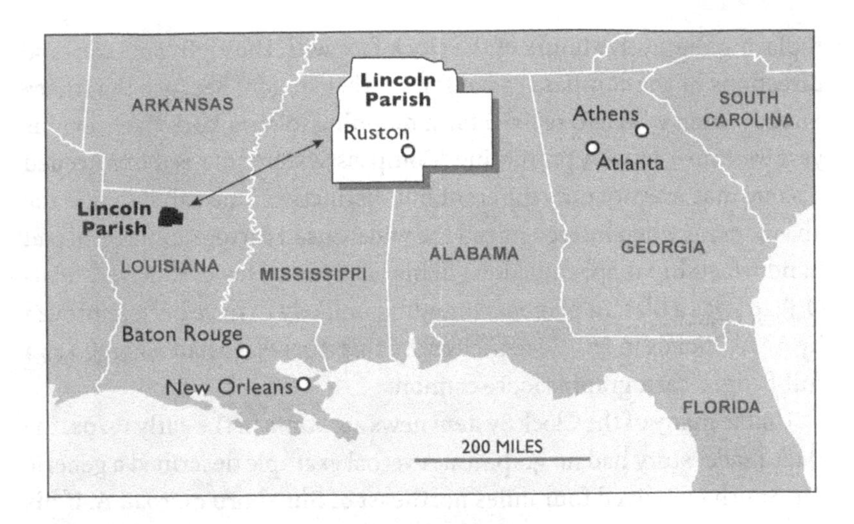

FIGURE 39. Ruston (inset at top center) was the county seat of Lincoln Parish, Louisiana, several states west of the AAA's Southern Division headquarters, in Athens, Georgia. Compiled by the author.

about 550 miles west of Athens, Georgia, and 170 miles northwest of the state capital Baton Rouge (fig. 39).

The initial news report originated in Ruston, the county seat of Lincoln Parish. On Wednesday, February 3, 1937, the *Ruston Daily Leader* announced that a mapping system soon to be introduced in the parish would "revolutionize the [country's] rural postal delivery system."[24] The previous evening "J. B. Plato, representing the U. S. Department of Agriculture," had explained the project at a meeting of the local chamber of commerce. The secretary of agriculture had personally selected Lincoln Parish for this "initial experiment in map making," and the project was a collaboration of the USDA, the Lincoln Parish Farm Bureau, and agricultural extension officials at the Louisiana State University in Baton Rouge. Based on a technique Plato had used "in several northern states, years ago, where it had proven commercially successful and desirable for use in many varied businesses," the "Plat-O-Plan" map would "give a house number to every dwelling in the parish, no matter how small or how remote from the county seat."

The news story included an abbreviated version of Plato's "lost sale" narrative that mentioned neither the inventor's watch nor the phrase "Clock System." Although direction was still very much a part of a rural address, replacing the twelve hours of the clock face with the eight sub-cardinal directions of the compass called for another slogan. Because the Ithaca businessmen who had revived farm mapping in New York the previous year were now actively promoting "Compass System" maps, Plato needed a name that was not only different but distinctive. One can imagine the aha moment when he recognized the widely used cartographic term *plat* (landownership map) embedded in his last name. However clever, "Plat-O-Plan" was a blatant personal branding unlikely to have been approved by AAA officials in DC. Nonetheless, by this time Plato had emerged as a full-fledged cartographic loose cannon.

Unlike many of the Clock System news accounts in the early 1920s, the *Daily Leader* story had no graphics. A verbal example described a generic Mr. Smith who lived four miles northeast of Simsboro on road A. If his distance or sequence number along that route were 250, his Plat-O-Plan address would be 250 A Simsboro NE 4.

Among other details, the article mentioned the inch-to-the-mile map scale, the directory listing residents alphabetically by last name, and the metal number plates, which "will probably be manufactured by the penitentiary," the go-to supplier of license plates. In addition to providing each household's map address, the directory would identify the occupant not only as an owner or a tenant but also as white or "colored"—a distinction never deemed relevant in New York or Ohio.

During the previous two months Plato had been using air photos to make a base map of roads, which were then driven by a field crew that marked every dwelling with a small dot, appropriately positioned on the left or right side of the road. The field crew had been recruited by the "parish agent" (county agent), identified as an active supporter of the project. The aerial photography had been flown for the Soil Conservation Service, whose field scientists used the imagery to identify boundaries between soil categories.

The second news article was distributed by the Central Press Association, a news syndicate that provided feature articles to local newspapers. Unlike breaking news, which lost value if not published shortly after an event occurred, feature stories were mailed to subscribing newspapers as stereo mats: paper-mâché or cardboard molds that could be converted into thin metal printing blocks in a "flat casting box" by adding molten lead.[25] An editor could assemble the layout for an entire page from printing blocks produced individually from stereo mats for ads and syndicate content. I

Government Begins Huge Task Of Numbering Isolated Farms

Louisiana Parish Sets Example For Remainder of Nation.

By K. F. HEWINS.
Central Press Correspondent

Ruston, La.—In the first project of its kind ever sponsored by federal government, Lincoln parish (county), in north Louisiana, has been surveyed and mapped for the preparation of a farm-numbering directory, designed to make the finding of a rural residence as easy as the locating of a city address.

The work has been in charge of J. B. Plato, now of Washington, D. C., but formerly a farmer in Colorado, who conceived the idea of such a directory 20 years ago, and recently interested the United States department of agriculture, which employed him.

That necessity mothered the original conception of the farm-directory system is shown by the explanation given by Plato in his office here, after the preliminary part of the work was done.

Handicap.

"While I was engaged in farming near Fort Collins, Colo., about 20

Twenty years ago J. R. Plato, above, conceived the idea of numbering farm houses. Now the U. S. department of agriculture has employed him to put such a directory plan into execution, Lincoln parish, Louisiana.

FIGURE 40. The second article describing the "Plat-O-Plan" map, distributed by the Central Press Association, included a photo of Plato. Reproduced with permission of the *Quad-City Times*, Davenport, Iowa. Image originally published as K. F. Hewins, "Government Begins Huge Task of Numbering Isolated Farms," *Quad-City Times* (Davenport, Iowa), February 12, 1937, 14.

found four examples of this second article in newspapers published between February 11 and 19.[26] Identical in wording—the stereo mat is the origin of the word *stereotype*—they all included an image that was a composite of an aerial photo, a drawing of a man looking at a Plat-O-Plan address, and a headshot of Plato himself (fig. 40).

The article reflects Plato's skill as a storyteller not averse to hyperbole.[27] The idea for a coordinated map and directory that made "the finding of a rural residence as easy as the locating of a city address" had been worked out twenty years ago by a Colorado farmer frustrated by the lack of a "real address" to use in advertising some Guernsey cows he wanted to sell. Although the scheme had been used for maps and farm directories in New York and Ohio, the Lincoln Parish map was noteworthy as "the first project of its kind ever sponsored by the federal government." Plato implied that his pilot project, "which is expected to be the beginning of a standardized system to be adopted on a wide scale, perhaps throughout the United States," had the support of Secretary of Agriculture Henry A. Wallace. The Secretary had selected Lincoln Parish because the county agent, who oversaw a cotton surveillance program, had acquired a complete set of aerial photos, equally useful for soils mapping.

The reporter attributed the unusual title "Plat-o-Plan Map of Lincoln Parish, Louisiana" to Plato's penchant for "catchy" phrases such as "Putting the Farmer on the Map" and "Taking the 'Far' Out of Farming." Field work was nearly complete, and the USDA had "withheld press reports of Plato's activities until the preliminary work was done." The map, which was being redrawn in Baton Rouge, would be printed by the agricultural extension service at Louisiana State University in cooperation with the USDA and the local farm bureau. It would be published in two sizes: a large format for office use and "a smaller size for individual use." No mention of pricing, free distribution, or a marketing plan. Unlike Plato's earlier projects, there was no provision for advertising—an unseemly prospect for a federal publication. Although the agency could distribute copies free or charge a small fee, publication depended on a line item in the USDA budget.

I found the third article in seven newspapers—there could be others—all published on the last day in February in widely dispersed locations such as Baton Rouge, where drafting work on the map was underway, and Rochester and Utica in Upstate New York, where many readers might have recalled Plato's county maps.[28] Simultaneous publication and identical wording reflect distribution by the Associated Press, which sent news stories electronically to subscribing newspapers.

Oddly, the lead paragraph began by mentioning a tornado that had destroyed houses and fences in Lincoln Parish the previous week—a poorly conceived journalistic hook insofar as the map's role in reconstruction would be minimal at best.[29] Quickly moving past the irrelevant tornado, the article was not only shorter than its predecessors but comparatively bland. It identified Plato as the project's supervisor, attributed the idea to his frustration as a Colorado farmer trying to sell a Guernsey cow, used a fictitious farmer living four miles west of Simsboro in its example of a "real address," and listed the map, directory, and number plate as the key elements of a system particularly useful for locating farms "off the main highway." Although a small "pocket" version would complement the large "office" maps, the story was silent about a release date or distribution plans. Nothing about the Secretary of Agriculture, and no mention of "Plat-O-Plan."

Publication of the map would have warranted a modicum of fanfare, but I found nothing, not even in the *Ruston Daily Leader*. The project never became a model for Plat-O-Plan maps of other parishes in Louisiana or elsewhere in the South. Although Plato apparently continued to make maps for the AAA and provide technical advice, his blatant self-promotion must surely have become repugnant to civil servants trying to stabilize agricultural prices. That October a final notice of termination marked the end of his career as a federal mapmaker.[30]

8 CAMP PLATO PLACE

Plato was not only ready to retire but content to remain in the neighborhood he chose when he moved to Washington in 1933. The three maps in figure 41 describe Plato's home in Forestville, Maryland, just outside the District in Prince George's County. The upper map was extracted from a 1923 road map drawn by the USGS. Although its horizontal extent is slightly less than ten miles, it shows Forestville as a significant commute from downtown Washington (on the left), where Plato had an office at the Census Bureau. It also shows the Benning neighborhood (top center below the G in Washington). Recall that the 1935 city directory reported two addresses for Plato: Benning, which is less than half the distance from downtown and adequately served by public transportation, and Forestville, well beyond the urban fringe. As the 1936 directory revealed, Plato chose the more sylvan Forestville, which was also a more troublesome commute. His Civil Service file in the National Archives includes a December 12, 1934, memo describing the combined impacts of distance, his erratic series of short-term appointments, and a sluggish government payroll.

> Dear Mr. Kile:
>
> The appointment of Mr. John B. Plato was made effective Wednesday, December 12.
>
> Mr. Plato has been working since October 22 on a special job for the Secretary of Agriculture, thinking that his appointment would be effective before this. We are requesting that you have the payroll prepared as soon as possible after this notification comes through because of the fact that he has run out of money and has no way of

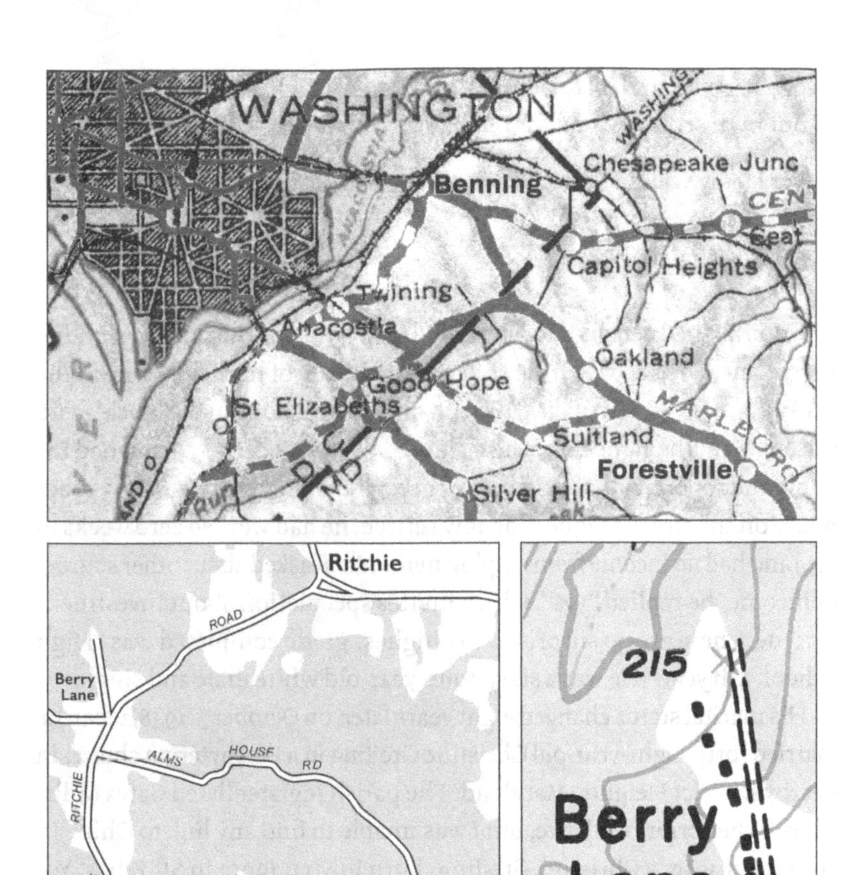

FIGURE 41. This trio of maps relates Forestville to downtown Washington (upper), the Berry Lane neighborhood to Forestville (lower left), and houses along Berry Lane (lower right) to nearby woodland, represented by a light gray tint. Upper Excerpt from *Road Map: Washington and Vicinity, Maryland—Virginia—District of Columbia, 1:250,000* (US Geological Survey, 1923). Digital image from the Library of Congress, http://hdl.loc.gov/loc.gmd/g3851p.ct004824. Selected features at lower left redrawn from US Geological Survey, *Upper Marlboro, Maryland* (quadrangle map), 1:31,680, 7.5-minute series, 1944. Lower right is a substantial enlargement of the Berry Lane portion of the Upper Marlboro topographic map.

getting back and forth between his home, which is quite a distance out in the country, and the office.

Very truly yours,
Wm. T. Wolfrey, Jr.
For Mr. Thatcher[1]

Unpublished census data located Plato's house at the northern end of Berry Lane, farther back from the street than any of his neighbors on the USGS map (fig. 41, lower left).[2] As with all his neighbors on the block, Plato was living in the same house he occupied five years earlier. He owned his home and reported a value ($10,000) greater than that listed for any other house on the street. Unquestionably retired, he had worked zero weeks in 1939 and had no income from employment. When asked about other sources of income, he replied "yes," which invites speculation about investment income. There are no surprises: his highest grade completed was "High School, 4th year." He was a sixty-four-year-old white male and single.

His marital status changed eight years later, on October 3, 1948, when he married forty-eight-year-old Christine Greiling in a Presbyterian church in nearby District Heights, Maryland. The parish register listed Gates Mills, Ohio, as her prior residence, but I was unable to find any link to Ohio.[3] In November 1939, a Christina Greiling, born July 14, 1900, in St. Vrbas, Yogoslavia [sic], had filed a Declaration of Intent to become a United States citizen; a resident of New York City, she had entered the country from Canada in 1937.[4] She reported July 10 as her birthday when she filed for Social Security benefits in 1965, but the Social Security Administration had apparently destroyed or lost her original application, which would have listed a place of birth.[5] When she died on April 2, 1991, at age 90, her will explicitly ruled out an obituary.

John lived nearly as long, but when he died on June 29, 1966, at age 89, multiple encomiums marked his passing. Probate records reveal that Christine, as his sole heir, received liquid assets worth $2,894, fourteen acres of land appraised at $63,000, and improvements (essentially a brick two-bedroom house with a carport and no basement) appraised at $15,500.[6] When Christine died a quarter century later, the house and land had already been sold, possibly to pay nursing home expenses, and specific bequests for items like a crocheted bedspread, a TV/stereo combo, and a sewing machine were no longer possible because they had been given away, possibly to the intended recipients.[7] Bank accounts worth more than $77,000 were held jointly by her executor (also known as a personal representative), presumably to pay day-to-day expenses.[8] According to probate records, the Platos' modest accumulation of wealth was further reduced by burial expenses and an inheritance tax of $3,875.

Newspaper stories indicate that Plato started purchasing land near his Berry Lane residence well before he retired. By the late 1940s and 1950s, newspaper references to the "John B. Plato estate" suggest that his acreage was contiguous or nearly so.[9] Always stated in round numbers, his holdings ranged between seventy and ninety acres. The *Washington Evening Star,* which ran a short item about his Japanese iris and their "180 blooms," mentioned his "70-acre backyard."[10] By contrast, a *Washington Post* story about the "84-year-old conservationist" who abhorred encroachment by shopping centers and apartment buildings described his coming to the county in 1934 and buying "90 acres nobody wanted."[11] His interest in farming, a brief throwback to his Colorado days, quickly waned. "I used to raise tobacco and then pigs," he said. "Now I let the Girl Scouts and Campfire Girls use some of the land. They come from all over. You couldn't get them out of there if you wanted to."

The *Post* ran the first report on his new hobby in July 1947. A large photo juxtaposed Plato, a hat shielding his face, with a girl on a pony, while seven other girls waited turns. A headline proclaimed, "Children Confer Title, 'Uncle John,' on Man Who Provided Forestville Camping Ground."[12] In addition to opening the "ninety-acre tract" to the scouts, he built them a lake and network of forest trails.

By 1951 the day camp had a name—Plato Place—and a professional director supervising volunteers who oversaw 220 campers divided into fifteen units.[13] The following year the *Post* mentioned Plato Place as one of three sites serving Girl Scouts throughout the District.[14] Plato was given an official thank-you for letting the girls "explore, hike, and camp" at Camp Plato Place. A 1962 photo in the archives of the Girl Scout Council of the Nation's Capital (fig. 42) shows Plato sitting in a car during one of his visits to the camp.

By 1956 Plato had made separate two-and-a-half-acre donations to two different Girl Scout groups. This was just a preamble. In 1966 his obituary confirmed a donation of fifty acres of woodland with three artificial lakes.[15]

FIGURE 42. J. B. Plato at Camp Plato Place, 1962. Girl Scout Council of the Nation's Capital. Used with permission.

As Christine attested, "the land was his whole life; he wanted to keep it in its natural state."

The following April the Associated Press reported a sad ending to the heroic tale of Uncle John's generosity.[16] Extensive vandalism, which county police and camp officials attributed to an increasing local population of teenagers, had made the camp unworkable, mostly because of the prohibitive cost of fencing the perimeter of the large property. "Almost every piece of equipment left on the 50-acre site over the winter has suffered some damage," the report said. In search of a solution, Girl Scout officials arranged a "land swap" that created Camp Aquasco Farm, in the extreme southeastern part of the county. Except as a tiny note on the Girl Scouts online history archives, Camp Plato Place is gone.[17] The land was sufficiently valuable to sustain the swap, but Plato's name is gone. I found no reference to a plaque at Aquasco commemorating Uncle John's kind regard for nature and the girls. The original site is now part of Walker Mill Regional Park, a 500-acre county park with sports fields and picnic facilities, but mostly open space.[18]

A plaque of a different sort seems a convenient place to end this chapter. Both John and Christine were cremated, but according to Lynn Entwisle, whose father was Christine's executor, John's urn had been buried somewhere on the property, presumably in a flowerbed inside a circular

FIGURE 43. Burial plaque for John and Christine Plato, on the grounds of Forest Memorial United Methodist Church, in Forestville, Maryland, and online at Find-a-Grave.com. Black-and-white rendering of image from https://www.findagrave.com/memorial/144341693/john-b-plato. Reproduced with permission of the photographer, Michael Rounds.

driveway. Eager to fulfill Christine's wish that they be interred together and commemorated by a large bronze burial plaque, Willard Entwisle had tried in vain to unearth the urn more than two decades later.[19] Though the bronze burial plaque (fig. 43) located in the small cemetery behind the Forest Memorial United Methodist Church includes both their names and life spans, it presumably covers only Christine's cremains.

There's more. Recall Christine's apparent confusion about her birthday (July 10 or 14)? Perhaps she had a penchant for garbling the dates of vital events. Whatever the reason, when the job order for the burial plaque was written out—it's there in the probate file—Plato's birth year became 1877, not 1876.[20]

9 REMISSION

Perhaps the strongest evidence for the usefulness of Plato's rural address system is the success of the Ithaca firm that resumed production of rural directories in 1936 and over the next five years extended coverage to half of New York's counties, reaching eastward to the Massachusetts border and westward to Lake Erie. A larger staff with offices in the new First National Bank Building[1] supported this comparatively rapid expansion as well as the remapping of all but two of the counties mapped by Plato.[2] Although these earlier compilations were no doubt consulted, the new maps had sharper, more logical symbols and more precise delineations. This chapter looks at the staffing and operations of the new firm, Rural Directories, Inc., including changes in map design and content. It also explores the plausible effects of a decline in number of farms and competition from telephone directories, also supported largely by advertisers.

Although Plato's maps, as well as their replacements, treated location as an amalgam of distance and direction from a locally important central place, the new maps replaced the Clock System's twelve sectors with the Compass System's eight cardinal and sub-cardinal directions, which, in my experience, is more straightforward than directions based on the clockface. If you are an Air Force fighter pilot flying in formation, "enemy bomber at 2 o'clock" is instantly understandable because the group's forward trajectory assigns an instinctive, undisputed direction to 12 o'clock. By contrast, on the ground one must first figure out north and then apply Plato's convention of 12 o'clock as the sector between due north and thirty degrees east of north; 1 o'clock as the sector between thirty degrees and sixty degrees east of north; and so on. As a geographer, I believe we all

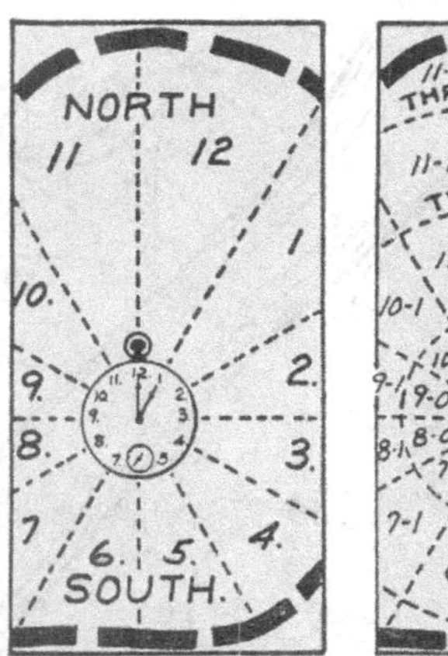

Shows "spaces" with their "Clock" No.

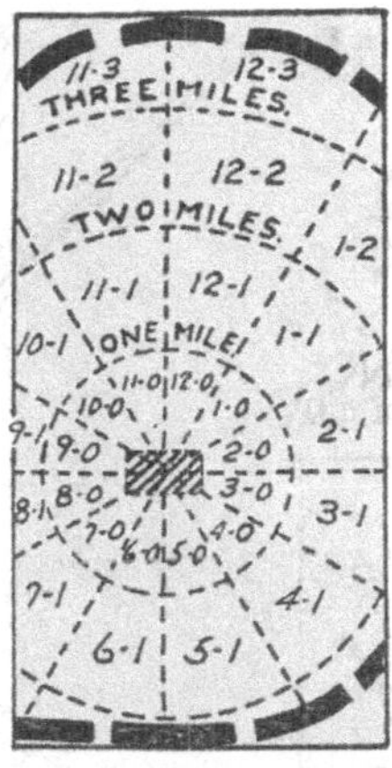

"Blocks" with their "Clock" and "Mile" No.

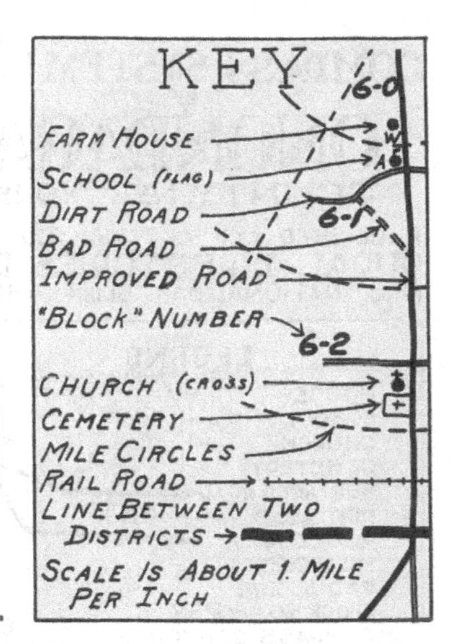

FIGURE 44. Explanatory diagrams and map key from J. B. Plato's 1927 map of Onondaga County. *"Clock System" Map of Onondaga County, New York*, series no. 16, 1:63,360 (Ithaca, NY: Index Map Co., 1927). Courtesy Geography and Map Division, Library of Congress.

should learn the cardinal directions around our homes, workplaces, and other familiar locations, but on the road it can get complicated.

The facsimile excerpts in figures 44 and 45 show how the Clock and Compass Systems differ in map content as well as address structure. The two elements to the left in figure 44 describe the basic direction-and-distance geometry of Plato's system, which partitions the district around a central place into blocks and assigns a unique letter to every farmhouse in the block. As the map key to their right indicates, a plain dot represents a farmhouse; a dot enhanced with a flag (pennant) represents a school; a dot enhanced with a cross represents a church; and a small rectangle with a cross in the middle is a cemetery. Compass System maps used similar symbols for these three features (fig. 45) but omitted railroads from the map key, even though these pervasive landmarks were often shown on the map by the traditional thin line with evenly spaced cross-ticks. Moreover, Compass System maps distinguished between only two kinds of road ("dirt" and "surfaced"), represented by thin and thick solid black lines,

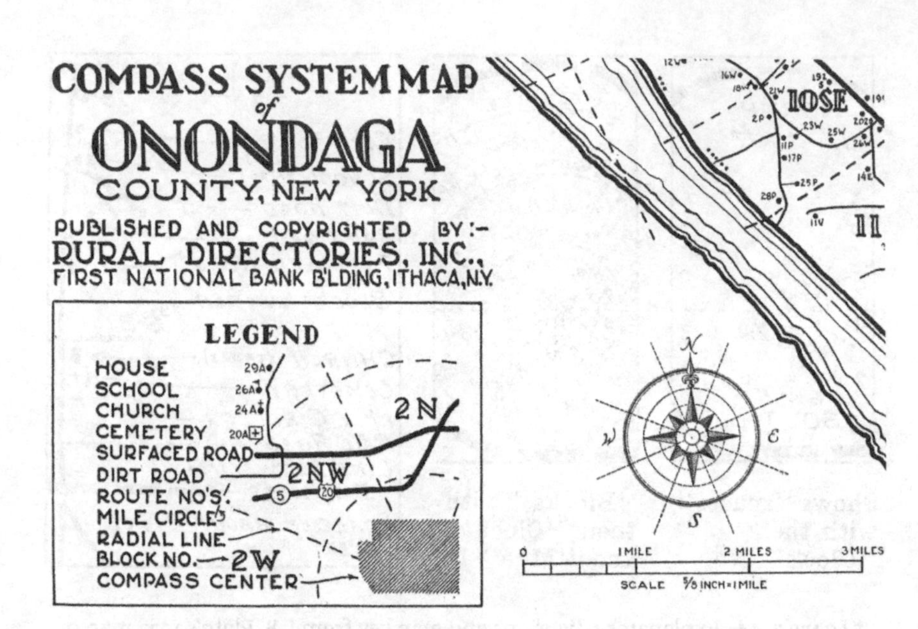

FIGURE 45. Map-key block from the 1936 Compass System map of Onondaga County. *Compass System Map of Onondaga County, New York*, 1:101,376 (Ithaca, NY: Rural Directories, 1936). Courtesy Maps and Cartographic Resources, Bird Library, Syracuse University.

whereas Plato's maps used parallel lines and thicker, solid-black lines as well as a third category ("bad") represented by a double-dashed line. In addition, with a third fewer radial lines, Compass maps were inherently less cluttered, allowing more space for two-part house labels in which a letter representing a particular road is preceded by an integer indicating the house's sequence along that road—essentially the same strategy Plato introduced in his final years at Ithaca (fig. 33) as well as in Ohio on his Farm Index Service maps (fig. 38).

Plato obviously was not using two-part house labels in 1927, when his Index Map Company published its Clock System map of Onondaga County, in the left-hand panel of figure 46. While some letters that are close in the alphabet are close within their block, others seem blatantly random—compare the sequence P, Q, R, and S in block 3-11, which is more systematic than the C, J, P, and S in block 3-9. In the left-hand panel, there's no consistent scheme to the assignment of letters: the only requirement is that a letter may not be used more than once in the same block. By contrast, note the increasing sequence 70G, 74G, 78G, 80G, 82G, 84G, 86G, 88G, 92G, 98G, and 102G in block 3N in the right-hand panel. Note too that as the road extends eastward toward Mycenae the sequence continues into two adjoining blocks as 108G, 112G, 120G, 124G, 136G, 142G, and 148G.

I composed the left- and right-hand elements of figure 46 to compare the Clock and Compass System maps published in 1927 and 1936 for Onondaga County, the first county mapped by Rural Directories, Inc. As similar in geographic scope as I could make them—both are anchored by the Madison–Onondaga county boundary on the east—these panels have been sized to reflect the printed maps' identical scale of one mile to the inch, exemplified by an identical one-inch spacing of consecutive circles.

Figure 46 reveals subtle differences in symbolization and content and not so subtle differences in legibility. In particular, note how the single letters in the left-hand panel from the Clock System map (1927) are larger and more readily readable than the smaller two-part house labels in the right-hand facsimile from the Compass System map (1936). In the lower-left corner of the left panel (1927) note the barely visible railroad symbol running from northwest to southeast through the V in Fayetteville. Toward the lower right is a zigzag double-dashed line representing a "bad" road with no houses. Then as now, the road was often clogged with snow in the winter. Note that how, going north, a jog to the right followed immediately by a jog to the left avoided the steeper gradient of a comparatively straight road that went directly up the hill and down the other side.

On the right side of the right panel (1936) a neat rectangle of cross-line shading delineates the hamlet of Mycenae, where non-farm residences are omitted. Prominent block labels stand out clearly, and lines representing highways are uniform in thickness, with unwavering straight segments as well as smoother curves than on the 1927 map. In addition to Green Lake and its smaller companion, the map includes a pair of highway numbers for State Route 290.

Block numbers reveal a fundamental structural difference in the extent of districts around central places. In the left panel a thick dashed line running northeasterly from the lower left corner is a district boundary separating blocks tributary to Syracuse (3-8, 3-9, 3-10, 3-11) from blocks in a 1 o'clock sector focused on the village of Manlius (1-2, 1-3), to the south. By contrast, almost the entire area within the right panel has addresses referenced to

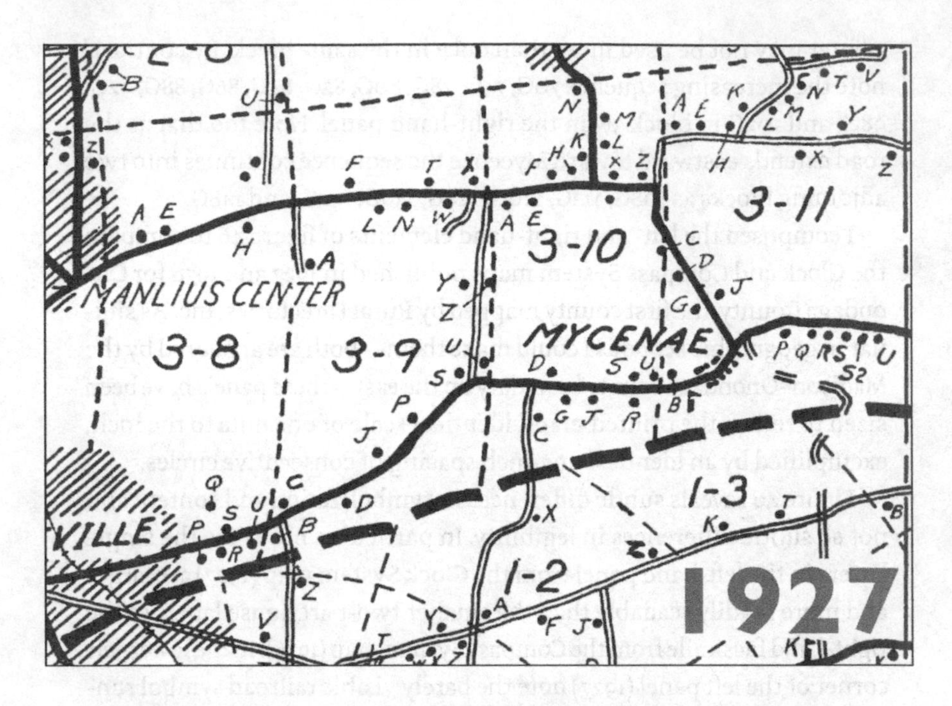

FIGURE 46. Juxtaposition of slightly reduced excerpts for roughly identical portions of the 1927 (left) and 1936 (right) maps of Onondaga County. Original size: 11.4 × 8.3 cm. Left: Extracted by author from *"Clock System" Map of Onondaga County, New York* (1927). Right: Extracted by author from *Compass System Map of Onondaga County, New York* (1936).

nearby Manlius in sectors extending outward to the north (2N, 3N, 4N, 5N) and northeast (3NE, 4NE, 5NE). In the extreme upper-left corner the sequence of large, closely spaced black dots is part of a boundary separating the Manlius and Syracuse districts. Only a tiny portion of the upper-left corner is still tributary to Syracuse. Clearly, the Compass System compilers did not endorse the shape and position of Plato's trade-area boundaries.

Discrepancies between the two maps underscore the need for revision. For example, house 11J in block 4N of the 1936 map is missing from block 3-8 of the 1927 map, perhaps because it had recently been built or discovered by a more careful field survey. Moreover, farmhouse positions on the newer map seem more precise—or less artificially spread out. The most prominent new feature on the 1936 map is the straight-line cut-off

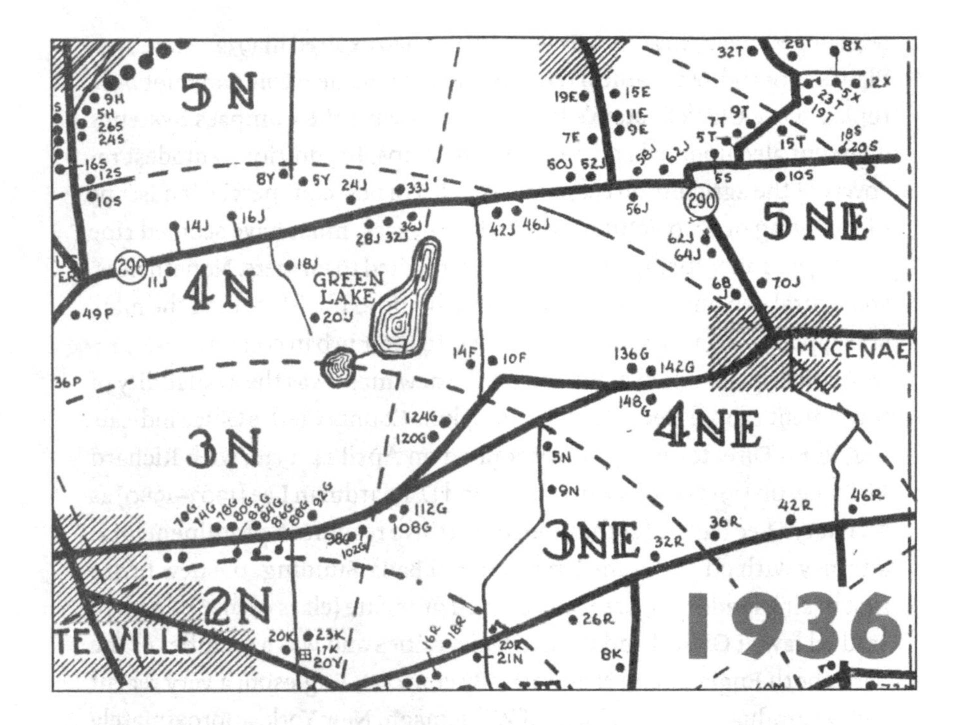

in blocks 4NE and 3NE on the paved road from Mycenae westward toward Fayetteville. Although scattered additions and subtractions confirm the need for a resurvey in the mid-1930s, the most prominent change is the map's redrafting: symbols as well as some positions look markedly different.

That said, as a business strategy, the most obvious change was the abandonment of the metal number plates, no doubt a hassle as well as a substantial bite out of Plato's bottom line. Rather than reissue new plates with each farm's new address—imagine the aggravation imposed on everyone involved—the new firm chose wisely to concentrate on its maps, directories, and advertising sales. Compass System addresses would be used mostly to find locations on the map, not as a substitute for signage on the landscape. Ostensibly a cost-saving measure, abandonment of the metal number plates also signified an erosion of Plato's map–directory–signage addressing system.

Why did this new firm emerge a half decade after Plato wound down the Index Map Company? Local people who recalled Plato's endeavor apparently recognized a publishing opportunity ripe for a reboot, and the only legal impediment, his Clock System patent, had expired in 1932.[3] Moreover, Plato's copyrights became irrelevant because the new company not only replaced the Clock System's twelve sectors with the Compass System's eight but also updated and redrafted the maps. In addition, a modest recovery of the agricultural economy, as well as a new commercial landscape of surviving or restructured service businesses, must have seemed ripe for targeted advertising in a publication pitched to farmers. Nonetheless, some rural consumers, not content with being explicitly put on the map, still appreciated a "real" address similar to their urban cousins.

Another plausible explanation for the new maps was the availability of investment capital. Records in the Tompkins County Clerk's Office indicate that Rural Directories was incorporated on April 14, 1936, with Richard M. Engleson (1915–1970) as president and D. Boardman Lee (1903–1980) as secretary.[4] Lee had by far the more impressive resume. A prominent local attorney with offices in the First National Bank Building, the new firm's first business address, Lee was a Cornell graduate (class of 1926) who had studied law at Oxford and Cornell Universities and was a member of the bar in both England and New York.[5] By contrast, Engleson, a very recent college graduate and a resident of Williamson, New York, approximately sixty-five miles northwest of Ithaca, was the son of Michael O. den Englesen, a wealthy produce dealer who died in 1931, when Richard was sixteen.[6] Rural Directories was incorporated three weeks shy of Engleson's twenty-first birthday, which suggests he might have been seeking a business opportunity in which to invest a recent or imminent inheritance. The only hint that he might have played an active role in the business is his listing in the Ithaca city directory for 1939, the only year when he was listed as a city resident.[7] The 1940 Census found him back in Williamson as general manager at a "farm machinery salesroom."[8]

Day-to-day operations at Rural Directories were overseen by H. Stilwell Brown (1905–1987), a 1927 graduate of Cornell's College of Architecture who had first-hand experience in farm mapping.[9] We know this because a May 1931 *Ithaca Journal* article reporting Brown's election as secretary-treasurer of the regional conference of American Business Clubs also mentioned that he was employed as a draftsman at the Index Map Company, then in a downward spiral.[10] In the early 1930s he had worked as both a salesman and an insurance inspector before joining Rural Directories as its executive vice president.[11] An October 1937 news article reported that he was also on the new firm's corporate board of directors.[12] His importance to the endeavor was underscored in April 1938, when the legal name of Rural Directories, Inc. was changed to Brown–Engleson Publishing Corporation.[13] Nonetheless, its maps and directories continued to be sold under the name Rural Directories until February 1938, when the company renamed itself Rural Surveys, Inc.[14] Engleson and Lee were still the corporation's president and secretary, but when Engleson moved to Ithaca and the city directory listed him as company president, Brown became its "managing editor."[15] After the company went out of business, Brown became sales manager at Cornell University's radio station WHCU.

The distinction between owners and operators was subtly apparent in an early May 1936 article on the front page of the *Ithaca Journal* below the headline "New Company Aims to Map Rural Areas."[16] The story identified Brown as the firm's executive vice president but didn't mention Engleson or Lee. Instead, it attributed the announcement to William Boyd, chairman of Ithaca Enterprises, Inc., newly chartered to foster local start-ups and attract existing businesses to the city.[17] The article cited Rural Directories, Inc. as "add[ing] another agricultural agency to the growing list of those having headquarters in Ithaca." Although the new firm was "using as a nucleus a system of mapping which was in operation here several years ago," its many changes included "a means of locating accurately and quickly the position on the map of any farmer in the county and indicating the best automobile routes to be used in reaching him"—recognition of increased automobility as well as a clear rationale for including state highway numbers.

The importance of advertising was apparent in the commitment to mail a free copy "to each farmer in the county." Complete coverage of this target market was essential to companies selling tractors, trucks, milking machines, milk coolers, poultry and dairy feeds, silage cutters, "antigen blood-tested leghorns," silos, farm structures of various types, and other agriculturally purposeful products advertised in the Onondaga County directory. No less prominent were ads for products and services useful to anyone, on or off a farm; examples include paint, furniture, "new and used radios," lumber, compressed cooking gas, plumbing and heating systems, washing machines, life insurance, funerals, and property financing and refinancing. Sellers offering shopping by telephone or mail as well as free delivery anticipated Amazon Prime by eight decades. In a four-fifth–page ad on the inside front cover, Dey Brothers Department Store in downtown

FIGURE 47. Folding over an extension of the directory cover and then stapling—twice each at top and bottom—provided a pocket on the inside back cover for storing the folded map. Scanned at a reduced size from directory with a trim size of 23 × 31 cm. *Rural Index Compass System Map and Localized Almanac, Jefferson County, N.Y.* (Ithaca, NY: Rural Surveys, 1939), inside back cover. Courtesy Cornell University Library, Rare and Manuscript Collections.

Syracuse promised "courteous and efficient service—and honest values in dependable merchandise" to the "6,300 Onondaga County residents listed in this directory." Making the customer feel important is a key part of any marketing pitch.

The postal service's role in full-market penetration was apparent in the bulk-mail permit number at the top of the front cover. The Cortland County directory, published in 1938, included the instruction "POSTMASTER: If undeliverable notify sender on Form 3547." Because the sender paid a small fee for each intended recipient who had moved, the address-change notification service was an inexpensive way of identifying entries requiring particular attention during the next field survey.

Two facts made the rural directory a reliable advertising vehicle: all farmers got one and most didn't throw them out. Although the map and its directory listings were the prime incentive for retention, the roster of advertisers, grouped by category in an index on the last page, was another reason for keeping the publication in a convenient, easy-to-find location in the home. And on the inside back cover a pocket for storing the folded map doubled as a half-page display ad for tractors and trucks (fig. 47). This example, from the Jefferson County directory, has separate lists of local sellers of steel-wheel tractors and pickup trucks.

No less important to some users were the articles, monthly calendars, almanac data, aphorisms, and lists of public officials scattered throughout the directory. In the Onondaga County directory, for instance, the "Almanac for March 1937" not only reported the approximate local times at which the sun and moon were to rise and set but also listed locally important historic events for every day in the month; examples include March 14, 1871, as the day "Thru trains on the D.L.&W.R.R. began" and March 26, 1879, as the day the "Messina Springs Hotel at Salina burned."[18] For each month a "Farm Record" box on the facing page listed important national and religious holidays and provided five blank spaces in which the farmer could write in noteworthy events or reminders. The farm record for May 1937 attributed the saying "But winter lingering, chills the lap of May" to poet Oliver Goldsmith.[19] Lists naming officials included the Onondaga Farm Bureau and the county's elected Board of Supervisors.[20] In addition, scattered boxes listed the titles and numbers of free bulletins that could be obtained simply by asking the county agent or sending a postcard to the New York State Colleges of Agriculture and Home Economics, at Cornell University. Bulletins were grouped into six categories; for example, "Agricultural Economics" and "Crops & Fruits." The twelve bulletins in the latter category included locally relevant titles such as "Growing on the muck soil of New York" and "Field mouse and rabbit control in New York orchards."[21]

The only other collaborators named by the *Ithaca Journal* were the five members of the advisory board. Figure 48, a facsimile of the list near the

BOARD *of* ADVISORS

C. C. CARPENTER
Prof. Botany, Syracuse University
EDWARD S. FOSTER
General Secretary, N. Y. State Farm
Bureau Fereration.
W. E. GEORGIA
N. Y. State Director, Rural Rehabilitation.
JOHN H. POST
Director of Welfare, City of Ithaca, N. Y.
DWIGHT SANDERSON
Prof. Rural Social Organization, N. Y.
State College of Agriculture.
HUGH J. WILLIAMS
Executive Director, Tompkins County
Development Assn.
PAUL WORK
Prof. Vegetable Crops, N. Y. State
College of Agriculture

The above board serves voluntarily
without compensation

FIGURE 48. Facsimile of list of Board of Advisors as shown on the inside title page of the Onondaga County directory (1936). *Onondaga County, N.Y. Rural Index and Almanac, with Map* (Ithaca, NY: Rural Directories, 1936), 3. Courtesy Maps and Cartographic Resources, Bird Library, Syracuse University.

front of the 1936 Onondaga County directory, indicates that two more members had been added before the directory was published later that year.[22] A note below the list reminded readers that, as unpaid volunteers, board members were not biased in favor of the publisher. Representing a broad range of academic fields and state and local agencies, the board might have met irregularly to suggest possible changes in design, content, or policy. In addition to recommending short articles, such as "Insect Control" and "Six Points in Chick Management" (included in the 1939 Albany County directory), advisors might have suggested helpful contacts in new areas the company planned to map.[23]

Although the *Ithaca Journal* reported that the new firm would have seven employees—"all Tompkins County residents except one" (Richard Engleson, presumably)—the article named only Brown. A few months later the Onondaga County directory's title page identified two more: Joseph A. Short (1909–1993), editor, and George Hoerner (1911–1981), cartographer.

Short, who graduated with a degree in drama from Ithaca College in 1935, was strongly interested in radio and community theater. After the business closed in 1940, he taught radio play production at his alma mater, where he headed the college alumni council in the mid-1950s. His principal job was at Cornell, where he served the university radio station WHCU at different times in various capacities, including staff writer, announcer, community liaison, production manager, and program director.[24] Hoerner shared Short's interest in the theater. Although he reported only one year of college to the 1940 Census, Ithaca College yearbooks indicate that after leaving Rural Directories around 1938 he worked at the college for a number of years as an "Instructor of Stage Craft" and eventually became chair of the drama department.[25] These later achievements attest to the competence and initiative of both Short and Hoerner.

Not all Compass System employees were named on the title page, at least not initially. The 1937 edition of the Ithaca city directory confirmed the *Ithaca Journal's* report that the company had six local employees.[26] In addition to Brown and Hoerner, the directory identified Carl L. Buchanan (1897–1971) as manager of advertising and sales promotion, Glenn E. Bullock (1913–1983) as "census taker," Helen S. Crispell as bookkeeper, and Thomas Kirk as "field census manager." Buchanan, who was active in the local advertising club, had been sales promotion manager of the Associated Gas and Electric Company before joining Rural Directories; he was promoted to vice president and sales manager, was made a director of the corporation, and remained with the mapping firm until its demise.[27] Whether Bullock ever participated in field operations is questionable. His name first appeared on the title page in 1936, listed below Hoerner as one of the firm's two cartographers, and from 1938 onward as its sole cartographer. He graduated from Cornell in 1933 with a BS in horticulture and, like many of his generation, he held a variety of jobs unrelated to his degree; after Rural Surveys closed in 1940, he worked as a draftsman for the New York State Electric and Gas Company and sold hearing aids for several years before becoming a postal clerk in 1948.[28] Crispell was listed as a mapping firm employee through the 1940 city directory, after which she was a clerk at another company. Kirk, who was listed in the 1937 and 1938 city directories, apparently left Ithaca after leaving Rural Directories.

Other personnel changes reflected on Compass System title pages include the hiring of Roger S. Reid as field supervisor in 1936 and the appointment of two corporate officers in 1937: Gilbert H. Sidenberg (1910–1944) as treasurer and Richard W. Sidenberg (1908–1986) as vice president. Although Reid's name remained on the title page from 1936 through early 1939, it never appeared in the city directory: hardly surprising insofar as his job did not require local residence. (Though the age range is appropriate, he was probably not the seventy-one-year-old Roger S. Reid who sold insurance in Syracuse and died in 1973—many people leave faint or confusing biographical footprints.[29]) By contrast, the Sidenberg brothers were Yale graduates who shared a house while working at Rural Surveys through 1940, and were later hailed as war heroes. Gilbert, who became an accountant in New York City before joining the Army in 1942, was killed in action in 1944 at the Battle of Bougainville in the Solomon Islands.[30] Richard, who was research director at radio station WHCU after Rural Surveys folded, received a Silver Star for "boldness, courage, and professional skill" during the early April 1945 assault on German positions near Magdeburg.[31] After the war he was general manager of the Roy H. Park publishing and printing firm in Ithaca before becoming regional director of Park's food distribution business.[32]

It's not clear who on the staff was in charge of media relations, a role Plato had managed single-handedly. Brown, as the operations executive, was almost always interviewed, but Buchanan was cited in an October 1937 *Ithaca Journal* story announcing the impending release of the Tompkins County edition.[33] Free copies were to be mailed to more than 3,800 rural residents. In a plea for local advertisers, Buchanan suggested the possibility of an annual edition. During the Compass System's half-decade run, only two counties warranted a second edition: Onondaga (1936 and 1938) and Cortland (1937 and 1940).

To explain the Compass System to *Journal* readers, the article juxtaposed excerpts from the Tompkins County Compass System map with an oblique air photo showing a specific farm and a portion of the spokes-and-circles grid around its local business center—the grid was inscribed as white dashed lines on the otherwise dark aerial image, and twin labels ("Here's the Farmhouse in the Picture" and "Here it is on the 'Compass System' Map") marked corresponding locations. Although the article's somber, splotchy halftoned aerial photograph printed poorly on newsprint, the concept was comparatively effective when reproduced with greater care on higher-quality paper in a rural directory (fig. 49). Overhead images,

Everyone realizes that "John Doe, R.F.D. 3, Farmville" indicates little regarding the exact location of John Doe's residence. To locate a certain individual "who lives on R.F.D. 3" invariably involves considerable guessing and searching. RURAL INDEX & MAP—with its "compass" address of each rural home outside the city and village limits and non-agricultural areas of Oswego County—now eliminates all such uncertainty. *See example as shown in box at bottom of this page.*
To the left: Aerial View taken near Taughannock Falls State Park, Trumansburg. (Camera pointed west.) Radial and circular lines have been superimposed on the photograph to illustrate the "Compass System" of locating rural homes.
Below: Section of Tompkins County "Compass System" Map for part of area shown in photograph above.
Here's the Farmhouse in the Picture–
2SE
2E
3E
Here it is on the "Compass System" Map.
This section of a typical county "Compass System" Map shows how the county is divided into zones, each having a city or village as its center. Radiating from this center are eight lines, dividing the surrounding area into zones designated by the four main and four intermediate points of the compass.
Concentric circles, a mile apart, divide the spokes into blocks which are designated as 1NW, 3SE, 2S, etc. ("3SE" indicates a block that is three miles south east of the center from which it radiates.)
the Compass IS THE KEY
LEGEND
HIGHWAYS
IMP. ROADS
DIRT ROADS
RAILROADS
HOUSES
COMPASS CENTER

FIGURE 49. Juxtaposed excerpts of an air photo and the Tompkins County map provided a concise graphic explanation of Compass System addresses. As shown in this example, from the 1938 Oswego County directory, text blocks were essential parts of the explanation. Reduced from 20.9 × 22.1 cm. *Rural Index and "Compass System" Map, Oswego County, N.Y.* (Ithaca, NY: Rural Directories, Inc., 1938), 2. Courtesy Maps and Cartographic Resources, Bird Library, Syracuse University.

used extensively in the late 1930s by the Soil Conservation Service and the Agricultural Adjustment Administration, added elements of authority and authenticity to the explanation.

Rural Directories was well situated geographically for high-quality reproduction. For photoengraving it relied on the Ithaca Engraving Co., located a block away; for printing it depended on the Norton Printing Co.,

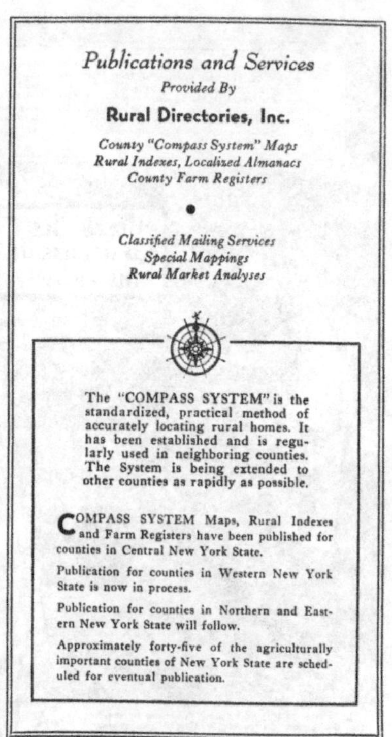

FIGURE 50. Central portions of the outside (left) and inside front covers (right) of jacket for special edition of the Tompkins County Compass System map, presented on February 24, 1938, at the annual Chamber of Commerce dinner. Courtesy Cornell University Library, Rare and Manuscript Collections.

two blocks to the east. In late 1938, when Rural Directories renamed itself Rural Surveys and moved a half block eastward, from the First National Bank Building to 147–151 East State Street (known to architectural historians as the Sprague Block), it was even closer to its printer.[34]

A special printing of the Tompkins County map underscored Rural Directories' close working relationship with its photoengraver and printer. A commemorative map naming all three firms on its cover (fig. 50, left) was presented to members of the Chamber of Commerce at their annual dinner in February 1938. The inside front cover (fig. 50, right) listed a variety of specialized products and services, including focused mailing lists covering all farms of a specific type (for example, dairy farms) across a multi-county region that grew larger with every new directory added. I have no information on the pricing or number of clients for these services, published as special editions of the county directories.

In addition to the "farm edition" distributed to all farmers, the company published two special editions, which were copyrighted separately. The "rural guide," with "Rural Register of All Farms" preceding the name of the county in its title, was identical to the county's "Special Classified Farm Register of All Farms," which also specified "which are Dairy, Fruit, Poultry Farms, etc."[35] Although main titles varied slightly from county to county, both included the subtitle "For Merchants, Manufacturers, and Professional Men." Both supplements promised every farmer's name, marital status, postal address, and "exact Compass System location" and claimed to be "unequalled for Circularizing, Locating Prospects, Demonstrations, Determining Credits, Cutting Delivery Costs, Exact Billing, and Prompt Collections."[36] Bare-bones, unpaginated lists of names and Compass System addresses, and devoid of advertising, these business versions included a folded copy of the county map in a pocket at the back, similar to the farm edition.[37] At the top of the front cover a unique stamped sequence number followed by a "Property of" blank for the name of the owner attested to a limited printing and carefully controlled circulation to business clients. The cover warned that copies were "sold under the implied terms that they will be used only in connection with the business of the purchaser" and that "publication in whole or in part will be prosecuted." Although a price was not shown, it was no doubt higher than the one or two dollars charged for individual copies of the farm edition, which lacked postal addresses.

An additional copyright was registered for the farm edition and sometimes a separate copyright for the map itself; all registrations reported an identical release date. Unlike Plato, who largely stopped registering copyrights in the mid-1920s, Rural Directories and Rural Surveys were conscientious registrants insofar as their listings in the *Catalog of Copyright Entries* match Cornell's extensive collection as well as the WorldCat database, which reflects library holdings. This impressive congruence endorses the quartet of maps (fig. 51) showing the yearly expansion of Compass System coverage as well as the month and day of publication of each set of maps, directories, and special editions. Each of these maps contrasts counties

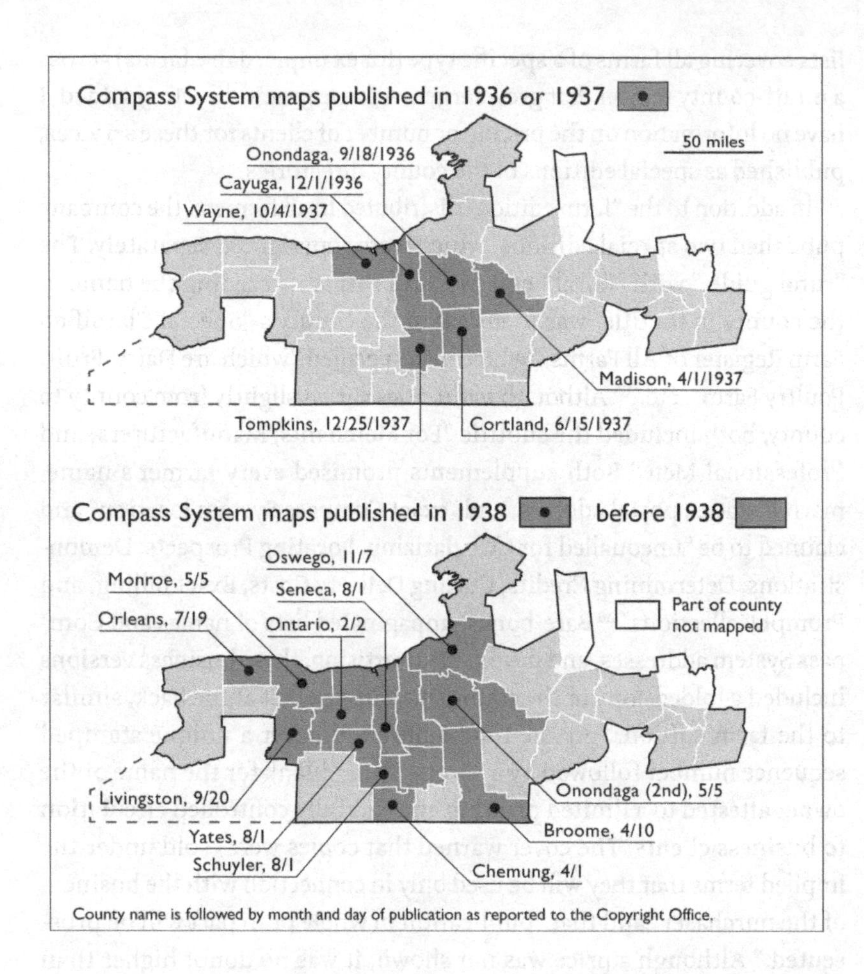

FIGURE 51. Expansion of Compass System coverage of New York counties, 1936–40, by year. The outer boundary of the coverage area encompasses all counties with either a Clock System or a Compass System map. Compiled by the author from the *Catalog of Copyright Entries*.

newly covered that year with those already covered. Dotted-line ellipses in the upper map in the right-hand panel represent combined maps and directories for Fulton and Montgomery Counties and for Schenectady County and the southern part of Saratoga County. Similarly, because farming was minimal farther north in the Adirondacks, only the southern part of Herkimer County was mapped.

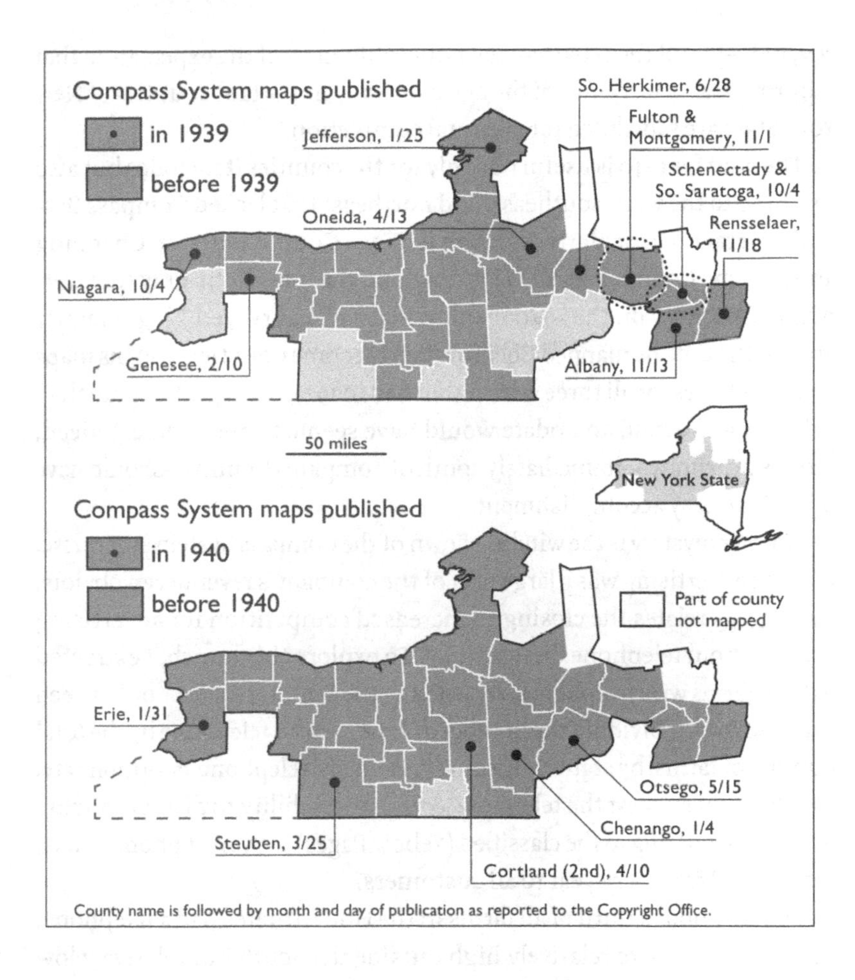

The maps reflect a coherent pattern of expansion described verbally on a special edition of the Tompkins County map presented in early 1938 at the annual Chamber of Commerce dinner (fig. 50, right-hand panel). By the end of 1937, coverage was essentially complete in a narrowly defined Central New York region. Work on counties in western New York was reliably reported as "now in process," and it seemed safe to predict that products for counties in northern and eastern parts of the state "will follow." Indeed, as a cartographic narrative, my four maps reflect steady progress on a rational expansion plan near the center of the state orchestrated by a well-staffed company. But when the last Compass System products were issued in mid-May 1940 for Otsego County, the list of thirty-two counties covered—two of them twice—fell noticeably short of the expectation that "approximately forty-five of the agriculturally important counties of New York State are scheduled for eventual completion."

The map for 1940 is useful not only for the counties it includes but also for those to the west, southeast, and northeast that lacked Compass System coverage. Particularly puzzling is Tioga County, between Chemung and Broome Counties on the 1938 map and the lone, light-gray (not-yet-mapped) county on the 1940 display. Plato had surveyed Tioga County around the time he mapped Chemung and Broome Counties, and his maps and directories for all three were published in 1922. Surely, after more than a decade and a half, an update would have seemed appropriate. Indeed, Tioga's proximity—immediately south of Tompkins County—should have made it an easy accomplishment.

A larger mystery is the winding down of the Compass System enterprise. Because advertising was a large part of the company's revenue, an obvious hypothesis relates the closing to increased competition for advertising revenue from telephone directories.[38] To explore this hunch, I extracted data on farms with telephones, reported in the *Census of Agriculture* between 1920 and 1964.[39] Dividing the number of farms with a telephone by the total number of farms, by state or by county, yielded a telephone-adoption rate, which might suggest the telephone companies' ability to divert Compass System advertising to the classified (Yellow Pages) section of phone books, distributed free every year to all customers.

By 1920 about a third of all farms in the United States had a telephone; adoption rates were relatively high outside the South but relatively low everywhere for tenant farmers. Installing poles and stringing wire was comparatively expensive in rural areas, where customers were markedly farther apart than in cities. Rural service was typically provided by a locally owned company, or exchange, which served multiple customers on simple circuits called party lines.[40] Each customer on the same party line was assigned a specific series of long or short rings: one short ring for Jones, one long ring for Smith, one short and one long ring for Brown, and so forth. A customer who wanted to place a call would pick up the earpiece and turn the crank of a small generator, which sent a signal to get the attention of the human operator in charge of the switchboard.[41] If Jones wanted to speak to Brown, the operator would ring the circuit two times—one short and one long—alerting anyone near the phone at the Brown residence to pick

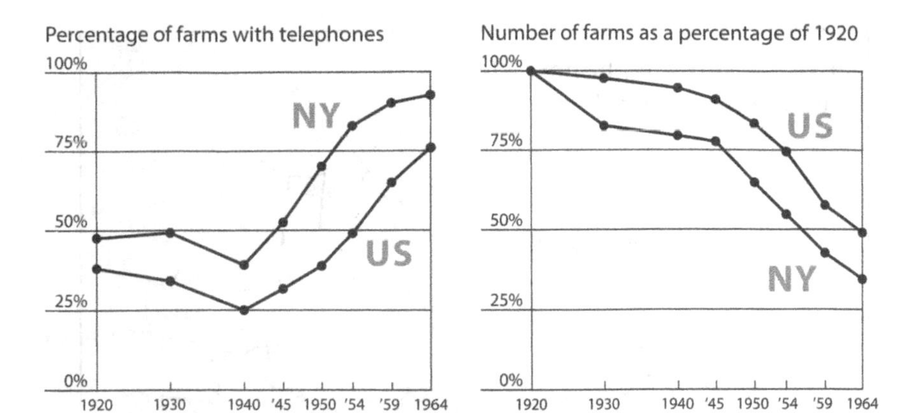

FIGURE 52. Time-series graphs of the telephone-adoption rates for farms, New York and the United States, 1920–64 (left) and of the number of farms as a percentage of total farms in 1920, in New York and the United States (right). Compiled by the author from *Census of Agriculture*, various years.

up. The switchboard could also connect the circuit with a caller on another local circuit as well as with a caller on another exchange.

Calls were supposed to be brief, and privacy was a courtesy not rigorously observed. It was not uncommon for a customer wanting to place a call to find the party line in use, which also made the phone an irresistible plaything for children when adults were not looking. Oldsters eavesdropped too, probably in far greater numbers. As the left-hand chart in figure 52 shows, a New York farmer was more likely to have a phone than the average American farmer. Because of rising costs and poor service, many farmers disconnected during the 1920s and 1930s, when increased automobile and radio ownership demonstrated that the telephone was not the sole remedy for rural isolation. That proportionately fewer New York farmers had telephones in 1940 than in 1930 reflects a national trend as well as the Great Depression, and the dramatic recovery after 1940 was propelled in part, as the right-hand chart confirms, by an at least equally dramatic decline in number of farms. Marginal farmers, more likely to sell out or merely quit, accounted for many of the disconnections.

Curious about how farmers' telephone-adoption rates might reflect potentially ominous competition from the Yellow Pages, I mapped these rates for New York counties in 1940, the Compass System's final year. If

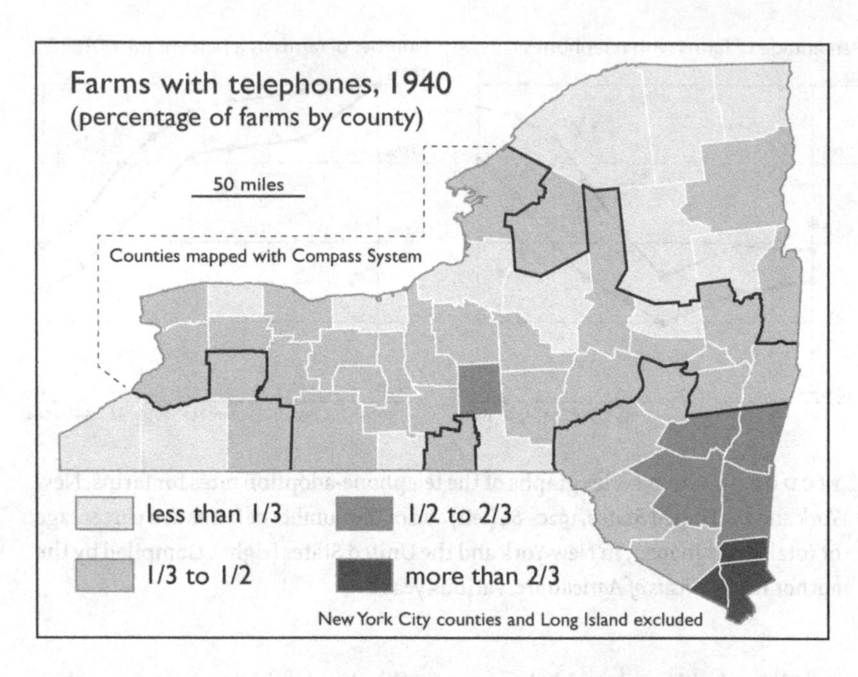

FIGURE 53. Telephone-adoption rate, Upstate New York, by county, 1940. A thick line highlights the counties wholly or partly covered by a Compass System map. Although the map's subtitle reflects use of percentage data, the fractional equivalents of 33.33%, 50%, and 66.66% in the map key invite more ready interpretation of the cut-points between categories. The map omits the five counties of New York City and the two additional counties on Long Island. Compiled by the author from *Census of Agriculture*, various years.

this hypothesis has any validity, higher-than-average rates in counties just outside the area with Compass System coverage might have discouraged further expansion, thereby hastening closure. The map in figure 53 offers limited support, insofar as the above-average rates in the four counties in the Hudson Valley south of Albany and Rensselaer Counties might suggest formidable competition from the Yellow Pages. No less significant are below-average rates in two unmapped counties in western New York and in five of the seven unmapped counties in northern New York. That relatively few farms in these counties had telephones probably reflects a weak agricultural economy and fewer farm-related businesses eager to advertise in a Compass System directory. Moreover, because these areas were relatively far from Ithaca, distance might have been a noteworthy impediment to recruiting institutional sponsors as well as advertisers.

A cursory comparison of telephone and Compass System directories for Onondaga County suggests another flaw in the Yellow Pages hypothesis. The county library has a paper copy of the 1936–37 phone book as well as later directories on microfilm. After comparing the 1936 Compass System directory with the 1936–37 and 1939–40 Yellow Pages, I concluded that the Compass System directory was helping a minority of advertisers pitch directly to the county's farmers. That two of Syracuse's large department stores advertised in the rural directory suggests that they considered farmers a customer segment worth cultivating. I found similarly selective Compass System advertising by a distinct minority of sellers in categories as diverse as banks, coal dealers, drugstores, electrical supplies, hardware dealers, insurance brokers, lumberyards, and paint stores. By contrast, a few business categories were wholly missing (even under a plausibly related rubric) from the Yellow Pages: Barn Equipment, Horse Dealer, and Silos. Perhaps the three silo contractors lacked phones or chose to rely on word of mouth.

Another factor that might have encouraged Engleson and Lee to shut down the company was an inability to raise the additional capital needed for continued expansion. In mid-October 1938 they filed a plan, approved by voting shareholders, to increase the official value of shares of capital stock from $25,000 to $50,000; shares had a par value of one dollar.[42] Five months later, however, they filed a proposal to reduce the 12,500 shares of common stock "issued and outstanding" to 2,000 shares of "Capital Stock A" and 2,500 shares of "Common Capital Stock A"—a significant devaluation of shareholders' investments.[43] The filing also spoke ominously of "receivership, liquidation, dissolution or winding-up of the corporation, whether voluntary or involuntary."

Fourteen months later the renamed Rural Surveys, Inc., published its final map, for Otsego County. The 1941 issue of the Ithaca city directory indicated that the USDA's Soil Conservation Service had taken over the company's space in the Sprague Block, and reported a new business address—the home of H. Stilwell Brown, now working for WHCU.[44] The 1942 issue dropped this last trace of a once-prominent local business.

10 POSTMORTEM

Eight decades without another encore for John Byron Plato's direction-plus-distance rural addressing system raises multiple questions. Was Plato's project a success or failure, or was it something more nuanced, such as right for the times—times that have changed? Did it meet a need, and if so, how pervasive was that need? What was geography's role in its success: a success defined narrowly in time and space? Was Plato one of many comparatively minor innovators, to be remembered in a history of cartography as little more than, to quote Macbeth via Shakespeare, "a walking shadow, a poor player that struts and frets his hour upon the stage and then is heard no more"?

My answer to this last question is a firm "yes," with the caveat that "then is heard no more" is surely—as this book attests—both premature and harsh. Now that I think of it, cartography has many "walking shadows," particularly in the age of digital mapping, which since it began in the mid-1960s rivaled Macbeth in leaving a stage strewn with corpses. Indeed, I am probably one of them, having authored numerous journal articles that enjoyed a brief run of acknowledgment as relevant literature before the inevitable citation taper. This is how cartography, like other sciences and technologies, moves forward by building upon or merely abandoning promising innovations, and leaving a residue to be swept up, with the cremains sometimes interred in a "history of" mausoleum. Someday, if I live sufficiently long, I might write a book about a few of them.

As conceded in chapter one, Plato was never properly memorialized in the twentieth-century volume of the *History of Cartography*, even though his Clock System experienced a modicum of success in map publishing and in helping beat back rural isolation. Its success was largely a response to several technological innovations, most notably the motor vehicle, which not only gave farmers a ready means for visiting towns, cities, and other farmers but also gave urban dwellers easier access to the surrounding countryside to buy farm products or sell services. Increased automobility after World War I reflected more affordable and reliable cars and trucks, continued encroachment by paved roads, and more readily available gasoline, all of which led to more frequent and longer trips as well as an increased demand for wayfinding maps and rural directories.

This demand was understandably greater in regions where farms were relatively small, dense, and profitable—regions such as Upstate New York, which was not only conveniently close to the large and growing megalopolitan market for dairy products, poultry, and fresh produce but also less easy to navigate than areas farther west, where the Public Land Survey System imposed its more straightforward, graph-paper geometry on local roads. Cornell University's enthusiastic academics enhanced a business environment in which merchants and manufacturers eager to sell to farmers created optimum conditions for the Clock System's success.

Other institutions provided valuable support. The agricultural extension service, with county agents linking farmers to land-grant colleges, gave Plato's business credibility as well as raw data and free publicity. And in the Clock System's early years, public schools supported the project by enlisting teachers and students in the labor-intensive task of collecting information and plotting locations, under the guise of an engaging geography lesson.

The Clock System was not the only solution to the problem of rural wayfinding. In setting up his company in Ithaca, Plato would surely have noticed another firm had been publishing rural directories for New York counties. Around 1913, Wilmer Atkinson (1840–1920), the Philadelphia publisher of *The Farm Journal*, a national monthly magazine that many farmers read, launched an auxiliary business that over the next half dozen years produced "illustrated farm directories" for over seventy counties in Michigan, Ohio, Indiana, and several Mid-Atlantic states, including New York.[1] Atkinson's directories were richer in information than Plato's insofar as farmers' listings not only distinguished renters from owners but mentioned the number of children in the household, the numbers of horses and cattle on the farm, the size of the property in acres, the telephone company (if any) to which the household subscribed, and whether

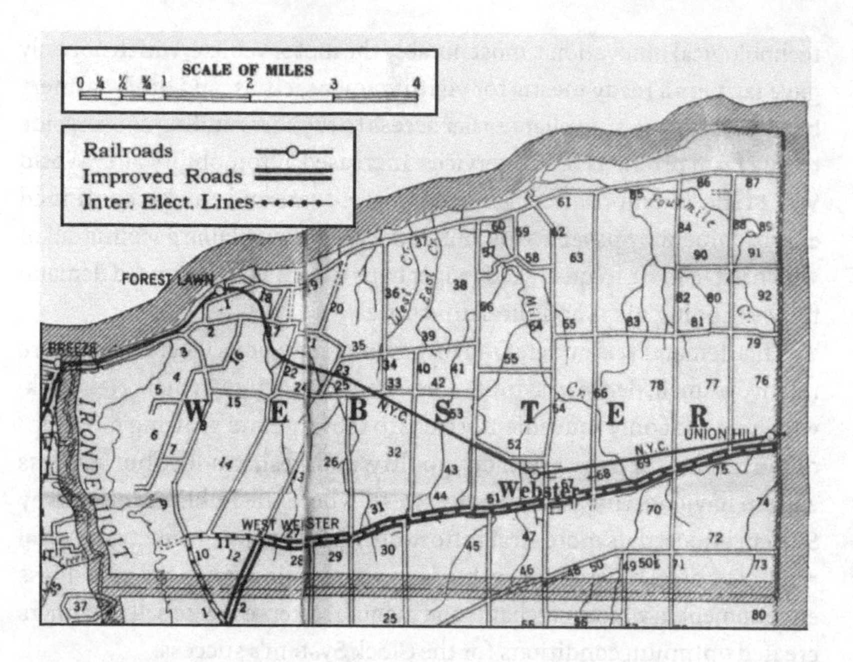

FIGURE 54. Town of Webster, New York, from the upper right-hand corner of the *Farm Journal Map of Monroe County, New York*, with excerpts for the bar scale and symbols identified in the map key. Excerpts from *Farm Journal Map of Monroe County, New York*, folded map accompanying the *Farm Journal Illustrated Rural Directory of Monroe County, New York* (Philadelphia: Wilmer Atkinson Co., 1918). Courtesy Central Library of Rochester and Monroe County, County Directories Collection.

the owner accommodated boarders. Their pages were smaller (6 × 9 inches, in contrast to the Clock System's 9 × 12 inches) but more numerous. Each directory had a subtitle in which a five-year period (for example, 1917–1922) promised durable information as well as eventual revision, but few if any were revised. *Farm Journal* lost interest in its rural directories after 1919, the year Plato published his inaugural rural guide for the Town of Ulysses.

Although Atkinson and Plato offered somewhat similar services, only Plato came close to providing a "real address." Instead of pinpointing individual farmsteads with a dot and label as Plato did, *Farm Journal* maps merely assigned residences to a numbered stretch of road between consecutive intersections (fig. 54). Mapmaking was attributed to the C. S. Hammond Company, a prominent New York publisher of maps and atlases.[2]

Although less detailed than Plato's, *Farm Journal*'s large folded maps were distinctively more polished.[3]

Plato could have used *Farm Journal* directories to confirm the completeness of his field surveys, to check the spelling of farmers' names, or to identify businesses that might buy an ad. However intriguing, this hunch is not confirmed by a comparison of his directories' geographic coverage with Atkinson's. Plato's activity more directly reflected proximity to Ithaca (fig. 30); he worked in only five of the eleven counties covered by a *Farm Journal* directory, all in western New York and all published in 1917 or 1918.[4] Distance mattered.

Was the half-decade encore in which the Compass replaced the Clock as the system's directional framework evidence of a real need, or can it be dismissed as a "dead cat bounce"? That the new firm mapped half of Upstate New York's counties most surely confirms a real need in the mid-1930s for a rural wayfinding aid. The company's principals were well positioned to recognize a distinct demand, which they met with a better organized, better financed, and overall more businesslike strategy than Plato's. The demand was there, at least in the late 1930s.

So why then, did this encore enterprise collapse? I can think of three reasons. First, competition for advertising—principally from telephone directories, which subscribers tended to keep—became more important as telephone hook-ups started to rebound (fig. 52, left). Second, the rural landscape and the business of agriculture were changing rapidly, as rural-to-urban migration drained off potential buyers and less viable farms were sold to more efficient neighbors or simply abandoned. Indeed, agriculture was in flux as mechanization made further inroads and dairy, poultry, and livestock farmers became less likely to grow their own feed.[5] Third, the Compass System's adaptable officers and employees apparently grasped other more stable or lucrative job opportunities as the country emerged from the Depression.

Advertising was probably the principal factor. My hunch is that selling ads was becoming more difficult in areas well removed from Ithaca. And having mapped New York's prime agricultural counties—at least those beyond the more immediate reach of New York City—the company perceived less enthusiasm among advertisers for revising its existing maps and directories. Paper was reasonably durable, after all, and many thrifty farmers might have kept their Compass System directories. Moreover, many businesses selling to farmers had to advertise anyway in a telephone directory, even if advertising only meant paying more for bold type. Although a few advertised on farm-oriented radio programs, local newspapers (both dailies or weeklies) were a more economical option for timely announcements of sales or new products.

Paradoxically, rural isolation was in retreat as migration was draining residents from the more remote rural counties. Factors behind this diminishing isolation include more paved roads, continued densification of the electrical and telephone networks, and commercial radio broadcasting, which offered entertainment as well as news, some pitched directly to farmers.[6]

Of course, the ultimate remedy for the "real address" problem Plato had sought to solve was the extension of house numbering into rural areas, accompanied by appropriate on-the-ground street-name signage. Institutions pushing for numerical addressing included emergency responders (police, fire, ambulance) as well as a broad range of delivery-oriented businesses such as fuel oil, Railway Express, and of course the Post Office itself, which recognized house numbers as more convenient and reliable than nonsequential box numbers for sorting mail, especially in regions like Upstate New York, where many city workers were choosing to commute from the outer suburbs and beyond.

Indeed, the Post Office took numeration even further by enhancing the mailing address with postal zone numbers, five-digit ZIP Codes, and ZIP+4 extensions, introduced in 1943, 1963, and 1983.[7] House numbering was firmly ensconced when the Census Bureau introduced the electronic Address Coding Guide (ACG) and Dual Independent Map Encoding (DIME) as strategies for expediting the 1960 and 1970 Census counts,[8] and emergency call centers began to link telephone numbers with street addresses. All this before GPS began providing turn-by-turn driving directions, and autonomous vehicles promised to take navigation out of the hands of human drivers. Whether printed or electronic, the map remains the quintessential wayfinding tool.

Ironically perhaps, a key invention for assigning house numbers originated in Ithaca, where Ralph Denman (1886–1975), a local civil engineer, became intrigued by strategies for rural house numbering. An advocate for the increased use of road names, which were easily extended beyond cities and villages into the surrounding countryside, Denman treated the

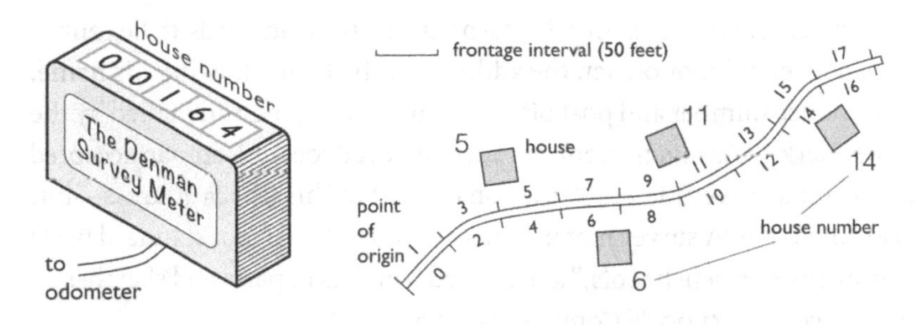

FIGURE 55. Principle of the Denman Survey Meter. A prescribed frontage interval divided a road into even- and odd-number segments (right) so that a counter advancing in increments of 2 (left) indicated house numbers for the right side of the road and adding 1 yielded house numbers for the left. Left: Redrawn from image in "Installation Diagram" in R. H. Denman, *Rural Planning and House Numbering* (Ithaca, NY: R. H. Denman, 1968), 12A. Right: Compiled by the author from discussion in Denman.

assignment of house numbers to rural dwellings as a mechanical process: simply start with 1 for the odd numbers on the left side of the road, increment by 2 for a fixed increment, and use a similar process for even numbers on the right side of the road. His "Century System" allocated a hundred numbers to the east side of a one-mile stretch of road, for a total of two hundred numbers on both sides. His approach was truly mechanical insofar as a counter connected to a car's odometer (fig. 55, left) clicked over a new number every fifty feet.[9] Because rural dwellings were typically at least one hundred feet apart, a so-called frontage interval of fifty feet was sufficiently robust to accommodate further subdivision and infilling (fig. 55, right).[10]

Although Denman was never awarded a patent, a drawing in his short, self-published book *Rural Planning and House Numbering* included the label "Pat. Ap'd For."[11] In the mid-1960s he contracted with several towns in and around Tompkins County to number residences along local roads. After making a suitably detailed map of roads and dwellings, he would drive outward from a road's origin, pencil in each resident's house number, and produce a street list with residents' names and addresses. Eager to promote house numbering, he enriched his book with a sample of a fill-in-the-blanks proposal a contractor could submit to local officials. Among the deliverables were "printed, postpaid Notification Cards to be sent to each householder or owner, the address on the front showing his name, road, house number and post office, ready to be signed and mailed by the Town Clerk."[12] As much an entrepreneur as an advocate, Denman accepted orders for a Survey Meter—"Price on request to" his Ithaca address—but also noted that "A survey meter . . . may be assembled from standard parts using common bench tools," and offered "sources of parts and sketches . . . upon receipt of $1.00."[13] Commendable outreach!

As an Ithaca resident, Denman could hardly ignore the prominent predecessor who had anchored georeferencing to town and village centers. Eager to distance himself from Plato, he began his book by listing no fewer than five reasons for the Clock System's decline. In addition to requiring both a map and a directory, which "were out of date in a short time," Plato's approach ignored well-known road names, assumed that users could quickly and reliably determine compass direction, and required special number plates for farmsteads or mailboxes.[14] Although these drawbacks did not thwart the Compass System start-up in 1936, Denman saw similar weaknesses in Plato's successors even though they identified roads with letters and dispensed with the cumbersome metal number plates. Cartographic innovations often come in waves, which break, spread, and dissipate.

In much the same way that John Byron Plato created a temporary cartographic fix for the farmer's lack of a "real address," extension of house numbering into the countryside created a renewed demand for paper wayfinding maps. The county-format indexed street map that became increasingly common in the 1950s and 1960s was an interim solution to a navigation problem largely solved by the in-vehicle satellite navigation system[15] and by the indexed street atlas (a paper solution where the street network had grown too large and intricate for a convenient folded single-sheet map).

John Byron Plato remains a minor figure in the history of cartography, but he is unquestionably distinctive. Although his rural directories can be helpful to genealogical researchers in Upstate New York and parts of Ohio, and collectors of transportation maps and quaint wayfinding tools might appreciate his depictions of past landscapes, his Clock System was less consequential than innovations at elite institutions like the USGS and Rand McNally. Even so, laic understandings of space and nonacademic interventions into spatial knowledge merit scholarly recognition. Particularly in Plato's case, for putting individual farms on the map as identifiable landmarks and recognizing farmsteads and other rural residences as elements of both social and wayfinding networks. He not only took a clever idea beyond a government-issued patent into the cartographic marketplace but promoted community identity in the face of disruptively reconfigured postal routes.

ACKNOWLEDGMENTS

In acquiring facsimile images and sources for tracing, I am greatly indebted to John Olson, maps and government documents librarian at Syracuse University (SU), and Ronda Frazier, archivist for Jefferson County, Colorado, who provided a multitude of details about Plato's lots in Semper, Colorado. Other librarians and archivists on my list of particularly helpful people include Evan Fay Earle, rare books and manuscripts librarian at Cornell University; Donna Eschenbrenner and Cynthia Kjellander-Cantu, at the Tompkins County History Center; John W. Hessler and Edward Redmond, in the Geography and Map Division of the US Library of Congress; Jenny Marie Johnson, map librarian at the University of Illinois at Urbana-Champaign; Annie Nelson and Brian Trembath, at the Denver Public Library; Dan Smith, in the local history department at the Onondaga County Public Library; and Luke Stegall, at the Public Library of Cincinnati and Hamilton County.

At SU's Bird Library, I relied on Darle Balfoort, in the map room; Natasha Cooper, collection development and analysis librarian; Carol Hamilton, interlibrary loans; Amanda Page, copyrights and open access librarian; Michael Pasqualoni, librarian for communications and public affairs; and Lydia Wasylenko, history and economics bibliographer.

For general information on copyright deposits, I am indebted to John Y. Cole, Library of Congress Historian. For information on postal maps, I am grateful to Virginia (Ginny) Mason, who wrote a thesis on the topic, and Stephen A. Kochersperger, in the history office at the US Postal Service.

For information on Plato's activities in Ohio in the early 1930s, I am especially grateful to Elaine Fletty, Wayne County Public Library. For information on his work in Louisiana, I am grateful to Derek Thompson, in the office of the Louisiana State Climatologist, and Suzy Ramm, editor of the Oregon State Grange Bulletin. For information on his land holdings in Maryland, I am indebted to Manuel Proctor, Assistant Chief Deputy, Information Technology Division, Prince George's County (MD) Register of Wills.

For information on Plato's support of the Girl Scouts in Prince George's County, Maryland, I am indebted to Ann Robertson, writer, editor, and Girl Scout historian in Washington, DC. Maryanne Lister, now living in the Shenandoah Valley, shared her memories of Plato from her days as a Girl Scout in Forestville, Maryland, and Lynn Entwisle, whose dad was Christine Plato's executor, provided information about the Platos' final years in Forestville.

In SU's geography department, Joe Stoll, staff cartographer and graphics guru extraordinaire, helped me navigate the idiosyncrasies of Adobe Illustrator and Photoshop. For day-to-day support on various fronts, I thank our dedicated office staff: Margie Johnson, Sarah Kondrk, and Debbie Toole. For network and systems support, I am grateful to the Maxwell School computer technology group, in particular Mike Cavallaro, Brian von Knoblauch, Tom Fazzio, Shannon Glennon, Daryl Olin, and Stan Ziemba. I also acknowledge the support of Christina Leigh Docteur, in SU's Office of Research; Caroline McMullin in the Office of Sponsored Programs; and Jill Ferguson in the Maxwell School.

Sincere thanks for continued support to my many allies in map history, especially Matthew Edney, Mary Pedley, Jörn Seemann, and Judy Tyner.

In preparing the manuscript I benefitted from the assistance of Marcia Hough and Victoria Lane. For professional advice on publishing, I am indebted to Lucas Church, Michael Gross, Bill Strachan, and David Wilk.

At the University of Iowa Press, director Jim McCoy's continued confidence in the project and insightful suggestions for restructuring the narrative were invaluable. I also appreciate the contributions of managing editor Susan Hill Newton, design and production manager Karen Copp, copyeditor Dan Forrest-Bank, and editorial assistant Jacob Roosa. As with the earlier books, I reaffirm a lasting appreciation of my wife Marge and daughter Jo Kerry.

NOTES

CHAPTER 1

1. Mileage data from *Railroad Ten-Year Trends, 2008–2017* (Washington, DC: Association of American Railroads, Economics and Finance Department, 2018), 17.

2. Interurban mileage peaked around 1918, following two growth spurts in the century's first decade. George W. Hilton and John F. Due, *The Electric Interurban Railways in America* (Stanford, CA: Stanford University Press, 1960), 2:208–9.

3. John B. Rae, *The American Automobile: A Brief History* (Chicago: University of Chicago Press, 1969), 7–11; and Reynold M. Wik, "Henry Ford's Tractors and American Agriculture," *Agricultural History* 38, no. 2 (1964): 79–86.

4. These estimates were extracted from a data graphic created by Nicholas Felton, on the *New York Times* graphics staff, and used to illustrate a Sunday Opinion article by W. Michael Cox and Richard Alm titled "You Are What You Spend" (February 10, 2008, 14). Felton's vertical axis was labeled "Percent of U.S. Households." His horizontal axis reported time in years from 1900 to 2005, and various curves described trends for telephone, electricity, auto, radio, stove, refrigerator, clothes washer, clothes dryer, air conditioning, dishwasher, microwave, color TV, VCR, cellphone, computer, and internet. I scanned Felton's graphic into a high-resolution image, densified his grid near where the curve crossed the vertical lines for 1910, 1920, and 1930, and estimated the percentages as 2 percent, 28 percent, and 59 percent.

5. Energy Information Administration, Office of Oil and Gas, US Department of Energy, *The Motor Gasoline Industry: Past, Present, and Future* (Washington, DC, 1991), esp. 1–15; Clay McShane, *The Automobile: A Chronology of Its Antecedents, Development, and Impact* (London and Chicago: Routledge, 2017), esp. 14, 21, 54; and Rae, *The American Automobile*, 49.

6. Theodore N. Beckman, "A Brief History of the Gasoline Service Station," *Journal of Historical Research in Marketing* 3, no. 2 (2011): 56–72; and John A. Jakle, "The American Gasoline Station, 1920 to 1970," *Journal of American Culture* 1 (1978): 520–42.

7. Peter Hugill, "Good Roads and the Automobile in the United States, 1880–1920," *Geographical Review* 72 (1982): 327–49.

8. Christopher Sweet, "A Comprehensive Bibliography of Nineteenth-Century Bicycling Periodicals," *American Periodicals: A Journal of History & Criticism* 29 (2019): 76–95, esp., 83–84.

9. James R. Akerman, "Road Mapping in Canada and the United States," in *The History of Cartography*, Vol. 6: *Cartography in the Twentieth Century*, ed. Mark Monmonier (Chicago: University of Chicago Press, 2015), (hereafter cited as *HC6*), 1339–50.

10. Richard F. Weingroff, "From Names to Numbers: The Origins of the U.S. Numbered Highway System," *AASHTO Quarterly Magazine* 76, no. 2 (1997); also online at https://www.fhwa.dot.gov/infrastructure/numbers.cfm.

11. James R. Akerman, "Twentieth-Century American Road Maps and the Making of a National Motorized Space," in *Cartographies of Travel and Navigation*, ed. James R. Akerman (Chicago: University of Chicago Press, 2006), 151–206, esp. 178–98; and Jakle, "The American Gasoline Station," 521–22.

12. Norman J. W. Thrower, "Cadastral Survey and County Atlases of the United States," *Cartographic Journal* 9, no. 1 (1972): 43–51; and Michael P. Conzen, "The County Landownership Map in America: Its Commercial Development and Social Transformation, 1814–1939," *Imago Mundi* 36 (1984): 9–31.

13. James A. Flatness, "Atlas: Subscription Atlas," *HC6*, 110–12.

14. Bates Harrington, *How 'Tis Done: A Thorough Ventilation of the Numerous Schemes Conducted by Wandering Canvassers; Together with the Various Advertising Dodges for the Swindling of the Public* (Syracuse, NY: W. I. Pattison, 1890), esp. 13–80.

15. Michael P. Conzen, "Landownership Maps and County Atlases," *Agricultural History* 58, no. 2 (1984): 118–22.

16. Clara Egli Le Gear (compiler), *United States Atlases: A List of National, State, County, City, and Regional Atlases in the Library of Congress* (Washington, DC: The Library of Congress, Reference Department, 1950), 213–22. Le Gear did not list any county directory published between 1860 and 1879 for Hamilton, Lewis, Madison, Putnam, and Seneca Counties. Her list included two directories that covered a pair of counties: Albany–Schenectady and Fulton–Montgomery.

17. Typical page size is from Harold Cramer, "Pennsylvania County Atlas Sampler," Historical Maps of Pennsylvania website, http://www.mapsofpa.com/atlaspage1.htm. Website last revised February 6, 2021.

18. Patrick H. McHaffie, "U. S. Geological Survey," *HC6*, 1659–66; and Roy Mullen, "Topographic Mapping: Topographic Mapping in the United States," *HC6*, 1555–59.

19. Robert C. Burtch, "Property Mapping: Property Mapping in Canada and the United States," *HC6*, 1194–200; Richard E. Dahlberg, "The Public Land Survey System: The American Rural Cadastre," *Computers, Environment and Urban Systems* 9, nos. 2–3 (1984): 145–53; Joel L. Morrison, "Coordinate Systems," *HC6*, 278–84; and Norman J. W. Thrower, "Cadastral Survey and County Atlases of the United States," *Cartographic Journal* 9 (1972): 43–51.

20. For a general history of Rural Free Delivery, see Wayne E. Fuller, *RFD: The Changing Face of Rural America* (Bloomington: Indiana University Press, 1964). For the cartographic history, see Virginia W. Mason, "The U.S. Post Office Department, Division of Topography: The Conception, Production, and Obsolescence of Postal Mapping in the United States" (M.A. thesis, geography, University of Wisconsin–Madison, 2002), 38–41.

21. Mason, "U.S. Post Office Department," 1, 20–33.

22. Mason, "U.S. Post Office Department," 35–37.

23. Lena Bedenbender Hecker, "The History of the Rural Free Mail Delivery in the United States" (MS thesis, State University of Iowa, 1920), 27–31. For a general history of Rural Free Delivery, see Wayne E. Fuller, *RFD: The Changing Face of Rural America* (Bloomington: Indiana University Press, 1964).

24. Mason, "U.S. Post Office Department," 31.

25. Mason, "U.S. Post Office Department," 53.

26. Bill Thornbrook and Sarah Thornbrook, *RFD Country!: Mailboxes and Post Offices of Rural America* (West Chester, PA: Schiffer Pub Limited, 1988).

27. Mason, "U.S. Post Office Department," 57–59.

28. Gerald Danzer, "City Maps and Plans," in *From Sea Charts to Satellite Images: Interpreting North American History through Maps*, ed. David Buisseret, 165–85 (Chicago: University of Chicago Press, 1990), 166.

29. William Wyckoff, "Cartography and Capitalism: George Clason and the Mapping of Western American Development, 1903–1931," *Journal of Historical Geography* 52 (2016): 48–60, esp. 53.

30. For a concise historical overview focused on the twentieth century, see Gerald A. Danzer, "Wayfinding and Travel Maps: Indexed Street Map," *HC6*, 1714–16. For a historically broader reference rich in images, see Nick Millea, *Street Mapping: An A to Z of Urban Cartography* (Oxford, UK: Bodleian Library, University of Oxford, 2003), esp. 10–17, 24–27. For further insights to the commercial street-mapping business in the 1970s, see Mark Monmonier, "Street Maps and Private-Sector Map Making: A Case Study of Two Firms,"

Cartographica 18, no. 3 (1981): 34–52 and Mark Monmonier, "The Geography of Urban Street Mapping in Pennsylvania: Recent Cartographic History," *Proceedings of the Pennsylvania Academy of Science* 54 (1980): 73–77.

31. *Clason's Guide Map of Denver, Colorado* (Denver: Clason Map Co., 1917). For discussion of the map's role in the larger enterprise, see Wyckoff, "Cartography and Capitalism," 53.

CHAPTER 2

1. For a history of Volume Six, see Mark Monmonier, "Brief Processual History of Volume 6," *HC6*, 1787–91.

2. Project newsletters can be retrieved at https://geography.wisc.edu/histcart/newsletter-archive/.

3. In writing this chapter, which carries the story through several other Denver residences, I relied on newspaper databases, Ancestry.com, public records, and city directories, which collectively offer insight to Plato's parents and grandparents as well as a hint of family money. Reconstructing Plato's family history was confounded by a lack of living relatives with a family album or memory box of old photos and news clippings in a closet or the attic.

4. Plato's father received a Bachelor of Laws degree from the University of Michigan in 1863, according to *University of Michigan Regents Proceedings, with Appendixes and Index, 1837–1864* (Ann Arbor, 1915), 1041. Whether he practiced law is not clear because he devoted much of his short career to selling lumber. According to the Cook County Death Index, he died on August 2, 1881, but the University of Michigan's *General Catalogue of Officers and Students, 1837–1911* (Ann Arbor, 1912, 372) reported July 28, 1881, as his date of death. A brief obituary in the *Parsons (KS) Weekly Sun* for August 5, 1881 (p. 1) noted that John B. Plato, "formerly a lumber dealer in this city, [who had] died at Geneva, Ill., a few days ago . . . was one of the early settlers of Parsons." Between June 5, 1873, and June 26, 1874, Denver's *Rocky Mountain News* ran a small daily display ad for his lumber business. The 1870 Census, accessed through Ancestry Library Edition, reported that on June 24, 1870, John B. (age 27, occupation illegible) and Hellen [sic] L. Plato (age 27, a "music teacher") were living in a house with nine other people in Manhattan, Kansas. Helen was part of that year's overcount because the 1870 Census manuscript covering Geneva included her in the household (age 27, a "schoolteacher" and "daughter-in-law") of her father-in-law (Judge Plato) along with Charles W. Plato, age 3, who (as I learned later from the judge's will) was the judge's adopted son. At the time Judge Plato was 61 and his wife, Almira, was 48.

5. An obituary for Plato's grandfather, the "Hon. Wm. B. Plato, of Geneva, Illinois," appeared in the (Denver) *Rocky Mountain News*, June 3, 1874, 1. "Judge

Plato had many friends in Colorado who will regret to hear of his death." He came to Illinois in 1837, settled in Aurora, read law with "Jo[seph?] Churchill," had a law practice in Batavia, went to the state legislature in 1858, was appointed "penitentiary commissioner" in 1859, served in the state senate, played an active role in Lincoln's campaign, and was "one of the early settlers of Greeley [Colorado]." Moreover, "his son, John B. Plato, esq., is one of the leading lumber merchants of our city."

6. Helen Frances Larrabee, a daughter of William M. Larrabee and a descendant of the Vermont Larrabees, was born in Chicago on October 4, 1842, and married John B. Plato in Geneva on December 27, 1864. Her husband was born in Aurora, Illinois, on September 16, 1842, and died in Geneva on July 28, 1881, "having had issue, three children." See G. T. (Gideon Tibbetts) Ridlon, Sr., *Saco Valley Settlements and Families: Historical, Biographical, Genealogical, Traditional, and Legendary* (Portland, ME, 1895), 799. Helen's obituary in the *Ithaca Journal*, March 5, 1931 (p. 5), mentions her burial "in the family plot in Geneva, Ill." Survivors included "one son, J. B. Plato of this city," two brothers, a sister-in-law ("Mrs. Helen Plato Wilbur of Geneva IL"), and a brother-in-law in New York City.

7. That grandfather, father, and inventor shared the middle name Byron might reflect the prominence of the celebrated aristocrat and poet Lord Byron, born George Gordon Noel Byron (1788–1824). See Mo Rocca, *Mobituaries: Great Lives Worth Reliving* (New York: Simon & Schuster, 2019), 309–13.

8. Kane County, Illinois, Wills and Probate Records, Box 37, file 2. Clerk of the Circuit Court (copies provided September 11 and 17, 2019). The handwritten will was not fully legible, but I believe this interpretation is correct. The probate records, mostly handwritten and apparently copied from microfilm, are generally consistent but somewhat confusing because of the sale of property in Geneva, Illinois, and lots in Greely, Colorado. According to executor's reports dated May 25, 1874, and May 26, 1903 [sic], the moneys distributed after payment of claims totaled either $3,037.33 or $3,597.57. For a brief biography, see "William B. Plato (deceased)," *Historical Encyclopedia of Illinois and History of Kane County*, eds. Newton Bateman and John S. Wilcox (Chicago: Munsell Publishing Co., 1904), 2:880.

9. Clerk of the Circuit Court, Kane County, Illinois, provided copies from microfilm of documents relating to the estate of John B. Plato, who died intestate "on or about" August 28, 1881. (Copies mailed on August 27, 2019.) Helen L. Plato, his wife, was appointed administratrix of his estate. She filed a final report with the Judge of the County Court on March 27, 1882. After claims were settled, the balance of the estate, which accrued to his wife, was $1,330.51. However, I found no probate records for Plato's mother or his paternal

grandmother. Helen Plato had been living in Tompkins County, New York, several years before her death in 1931; the Tomkins County Surrogate Court had no probate records. Records for Plato's grandmother, Almira or Elmira Plato (1815/1820–1895) were scanty at best—I am still uncertain of her birth year. A search of court index records in Kane County, Illinois, revealed nothing.

10. The court clerk's office in Prince George's County, Maryland, provided a copy of Plato's will and related probate records for estate no. RE-17023. (Copies emailed on August 15, 2019.) According to the report of his executrix, filed on March 6, 1967, the residue of his estate, all of which was willed to his wife, was valued at $61,485.00; it included 14 acres of land and a one-story, two-bedroom house with kitchen, living room, one bathroom, no basement. Because a 0.4-acre property in Semper, CO, was not included, and because his wife, Christine Plato, sold it about eight years after his death for $2,500, the Jefferson County, CO, archives also had a copy of his one-page will, a Jefferson County certification of authenticity, and a record of the sale and transfer deed in book 2480, pp. 115–18. (Copies emailed July 23, 2019.)

11. Denver city directories mentioned in this paragraph were accessed through Ancestry Library Edition, a searchable database with full-page scans for Denver directories published yearly in 1866, 1871, 1873–1889, 1891–1945, and 1947 and beyond.

12. Helen Plato first appeared in a Denver directory in 1891, when she lived at 2825 Gilpin Avenue, the home of her mother (J. B.'s grandmother), Mary M. Larrabee, and her brother Philip F. Larrabee. A 1921 fire destroyed most records from the 1890 Census, but archives for the 1900 enumeration place Helen and her son John at 2825 Gilpin, along with her mother Mary, Philip, Philip's wife (also Mary), her brother Frank, her mother's eighty-two-year-old sister Eliza Haight, and a thirty-two-year-old nurse Margaret Sullivan.

13. *Ballenger & Richards Twenty-sixth Annual Denver City Directory* (Denver, 1898), 670 and 893. The 1898 directory lists only his maternal grandmother (Mary M. Larrabee) and an uncle (Philip Larrabee) sharing the house at 2825 Gilpin, but the 1900 Census added another uncle, an aunt, a grandaunt, and a thirty-two-year-old nurse. See John B. Plato, 1900 census, Denver, Arapahoe County, Colorado, precinct 4, ward 8, supervisor's district 24, enumeration district 59, sheet 35, Ancestry.com (July 30, 2019).

14. For a description of the "Winter Course," see Charles Arthur Taylor, *A Brief History of Fifty Years of the Winter Short Courses, New York State College of Agriculture, Cornell University Ithaca, New York 1892–1942* (Ithaca, NY, April 25, 1942), which reports its author's credentials as "B. S., (Cornell '14), Professor in Extension Teaching, Cornell University."

15. *The Cornell University Register, May 1896*, 2nd ed. (Ithaca, NY: University Press of Andrus & Church, 1896), 141. In 1927, when the *Ithaca Journal* profiled his firm in the twentieth installment of its "Know Ithaca" series about local businesses, the second paragraph revealed, "Some twenty-five or thirty years ago when Cornell first started the winter courses for farm boys and girls there was one boy who came from Colorado to take the course. He had never lived on a farm and it was twenty years later before he had a farm of his own." "Know Ithaca: Index Map Company, Inc.," *Ithaca Journal*, October 12, 1927, 13. How Plato learned of the Cornell program is a mystery. I found no evidence that he had ever lived on a farm, or that a relative owned or had previously owned a farm.

16. *Cornell University Register, May 1896*, 264–65.

17. *The Cornellian*, vol. 28 (Ithaca, NY: Junior Class of Cornell University, 1896), 189.

18. "Agricultural Association," *Cornell Daily Sun*, March 4, 1896, 1.

19. Social note ("The second class of the Manual Training High School held its first class social . . ."), *Denver Post*, October 30, 1897, 5; Social note ("Friday evening the second class of the Manual Training school . . ."), *Denver Rocky Mountain News*, February 13, 1898, 20; "Joy Over the News: Soldiers at Camp Adams Receive the Welcome Tidings," *Denver Post*, May 9, 1898, 5; "Letters from Our Boys," *Colorado Transcript* (Golden, CO), October 5, 1898, 1; and Arthur C. Johnson, *Official History of the Operations of the First Colorado Infantry, U. S. V. in the Campaign in the Philippine Islands* (no publisher identified, 1899), 57. Plato's military career was marred by a 21-hour AWOL episode, from 6 p.m. June 29, 1899, to 3 p.m. June 30. According to his military service record in the National Archives, he was "found wandering along our lines without authority," arrested, tried, found guilty, and fined $2.00 (about four days' pay). He was discharged in San Francisco on July 31, 1899. National Archives and Records Administration, G11-539340059P, Pvt. John B. Plato Military Service Record, pp. 3–4, 6–7 (electronic delivery January 22, 2020).

20. Jim McNally, *Denver's Manual High School, 1876–2016: Bricklayers to Thunderbolts* (Denver, CO, 2016), 11.

21. "The First Colorado Regiment—Its Record in War and in Peace," *Colorado Springs Gazette*, September 13, 1899, 11–13; and "Formal and Informal" ("The reunion and banquet of the class of '99 of the Manual Training High school . . ."), *Denver Rocky Mountain News*, June 29, 1902, 20.

22. John C. Callahan, *The Fine Hardwood Veneer Industry in the United States: 1838–1990* (Lake Ann, MI: National Woodlands Publishing Company, 1990), xiii–xvi.

23. US Dept. of Agriculture, Forest Service, *Forest Products of the United States, 1906*, Forest Service Bulletin no. 77 (1908), 96–99.

24. David Denning, *The Art and Craft of Cabinet-Making: A Practical Handbook to the Construction of Cabinet Furniture, the Use of Tools, Formation of Joints, Hints on Designing and Setting Out Work, Veneering, Etc., Together with a Review of the Development of Furniture* (London: Whittaker & Co., 1891), esp. 4–5, 34–35, 41–45, 177, 221–38.

25. Building permit (1902) 8-1.00#1010.00 authorized an 18 × 29-foot structure with an estimated cost of $1,000 on lots 12–13 in block 22 in the Schinners subdivision. Page 303 of 483 in Denver Building Permits, Denver Public Library Digital Collections, http://digital.denverlibrary.org/cdm/fullbrowser/collection/p16079coll8/id/1527/rv/compoundobject/cpd/1708/rec/1.

26. The citywide base map was the *Map of Denver*, produced in 1900 and distributed by the Denver Chamber of Commerce and Board of Trade; downloaded from the digital collections of the Denver Public Library (http://digital.denverlibrary.org/cdm/maps/). The city directories were published for various years by Ballenger & Richards, of Denver, Colorado. The trio of large-scale atlases are *Baist's Real Estate Atlas of Surveys of Denver, Col.* (Philadelphia: G. Wm. Baist, 1905), and Sanborn Map Co., *Fire Insurance Maps of Denver, Colorado* (New York: Sanborn Map Co., 1903–4 and 1929–30).

27. For a historical overview of this type of large-scale map, see Ronald E. Grim, "Fire Insurance Map," *HC6*, 428–30.

28. For an interactive map, see Ryan Keeney, "Denver's Streetcar Legacy and Its Role in Neighborhood Walkability," capstone project, M.S. in Geographic Information Science, University of Denver, 2017, https://dugis.maps.arcgis.com/apps/MapSeries/index.html?appi=00a2d498a2ac4c58ad140ac306110213.

29. Brian Trembath, Western History and Genealogy Dept., Denver Public Library, email, August 12, 2019.

30. "Bachelor Makes Children's Playground of Back Yard," *Denver Post*, October 23, 1909, 10. I inferred that Plato initiated the playground in 1906 from the article's comment that his "tak[ing] the trouble" to develop the playground "has been done for the last two or three years."

31. Mathew Jancer, "The Amish Horse-Drawn Buggy Is More Tech-Forward Than You Think," *Popular Mechanics*, January 9, 2017, online only. https://www.popularmechanics.com/cars/car-technology/a24666/how-the-amish-build-a-buggy/.

32. Ballenger & Richards *Thirty-Fifth Annual Denver City Directory for 1907* (Denver: Ballenger & Richards, 1907), 1542.

33. J. B. Plato and W. H. Kilgore, "Hitching Attachment for Vehicles," US Patent 807,047, filed April 29, 1905, and issued December 6, 1905. The patent's title at the beginning of the text (but not above the illustration on the first page, shown here in figure 11) adds, "Said Kilgore Assignor to Said Plato."

34. See, for example, *Ballenger & Richards Thirty-First Annual Denver City Directory for 1903* (Denver: Ballenger & Richards, 1903), 655. The 1902 directory (p. 622) placed him in rented rooms at 2507 15th Street and also driving for Denver Towel Supply.

35. William H. Kilgore and Amanda Kilgore divorced in Denver on June 13, 1901, according to Ancestry.com, Colorado, Divorce Index, 1851–1985, and case 32756 in Historical Record Index, Colorado Archives, www.colorado.gov/pacific/archives/. Kilgore's history is inconclusive insofar as Ancestry.com reports the residence in Denver in the early 1900s (but rarely during the same year) of a Will Kilgore who might or might not be William H. Kilgore. The issue is difficult to resolve because records for Will and William are skimpy.

36. See, for example, *Ballenger & Richards Thirty-First Annual Denver City Directory for 1903*, 685.

37. *Ballenger & Richards Thirty-Third Annual Denver City Directory for 1905* (Denver: Ballenger & Richards, 1905) listed Rollandet's business address, "33 Jacobson Bldg. opp Post Office," on pp. 1321, 1364, 1378, 1445, 1477, and 1479 (sometimes in multiple listings on the same page), in addition to a full-page ad on 1478.

38. William J. Rankin, "The 'Person Skilled in the Art' Is Really Quite Conventional: U.S. Patent Drawings and the Persona of the Inventor, 1870–2005," in *Making and Unmaking Intellectual Property: Creative Production in Legal and Cultural Perspective*, eds. Mario Biagioli, Peter Jaszi, and Martha Woodmansee (Chicago: University of Chicago Press, 2011), 55–75.

39. The masculine pronoun is appropriate, insofar as the inventors I discussed in *Patents and Cartographic Inventions: A New Perspective for Map History* (Cham, Switzerland: Palgrave-Macmillan, 2017) were overwhelmingly male for most categories of map-related inventions and periods ending shortly after World War II. The notable exceptions were Ellen Fitz and Elizabeth Oram, prominent nineteenth-century pedagogues who patented globes.

40. An approved patent is assigned a sequence number once the examiner approves all changes, the inventor pays all fees, and the patent is officially awarded. Correspondence is kept on file indefinitely at the National Archives and is retrieved by patent (award) number. Although Plato and Kilgore's patent was issued on December 6, 1905, an official "notice of allowance," indicating a satisfactory vetting, was dated October 16, 1905, less than six months after its filing date, April 29, 1905, when the application arrived at the Patent Office.

41. Quotations from p. 3 of US Patent 807,047.

42. J. B. Plato, "Horse-Hitching Device," US Patent 823,964, filed June 21, 1905, and issued June 19, 1906.

43. J. B. Plato, "Hitching Device," US Patent 986,591, filed December 26, 1905, and issued March 14, 1911.

44. The case file in the National Archives for patent 986,591 indicates this sequence: rejected May 5, 1906—amended April 27, 1907; rejected June 7, 1907—amended May 21, 1908; rejected June 30, 1908—amended June 8, 1909; and rejected July 12, 1909—amended July 12, 1910.

45. To avoid typographic clutter, I suppressed "St." and "Ave." and shortened the business description "Veneers, Cabinet Woods" to "Veneers."

46. See pages 995 and 1511, respectively, in the general and business sections of *Ballenger & Richards Thirty-Fourth Annual Denver City Directory for 1906* (Denver: Ballenger & Richards, 1906).

47. I don't know whether he was a full or part owner of the lumberyard, in which he could have invested proceeds from the sale of the house on Williams—there is much about Plato we'll never know.

48. A reliable electric starter, which depended on a reliable motor, a reliable storage battery, and a reliable connection to the crankshaft, was not yet available. Curtis D. Anderson and Judy Anderson, *Electric and Hybrid Cars: A History*, 2nd ed. (Jefferson, NC: McFarland and Co., 2010), 15–17.

49. US Bureau of the Census, *Statistical History of the United States from Colonial Times to the Present* (Stamford, CT: Fairfield Publishers, 1965), Historical Statistics, Series Q 312, Motor-vehicle Factory Sales, Motor trucks and busses, 462. For additional insights, see Clay McShane and Joel Tarr, "The Decline of the Urban Horse in American Cities," *Journal of Transport History* 24, no. 2 (2003): 177–98; and Gijs P. A. Mom and David A. Kirsch, "Technologies in Tension: Horses, Electric Trucks, and the Motorization of American Cities, 1900–1925," *Technology and Culture* 42, no. 3 (2001): 489–518.

50. Jefferson County, Colorado, "Historic Land Records," book 169, 228, tracts 1 and 3, Semper Garden Tracts, Arthur McGuhan to John B. Plato, January 26, 1910, warranty deed. The warranty deed specified a cash payment of $10 plus "other valuable consideration," an apparent reference to the $790 mortgage, granted by the seller and secured by a trust deed, signed the same day. Jefferson County, Colorado, "Historic Land Records," book 168, 314.

51. Dawn Bunyak and others, *Westminster Selective Intensive Survey, Jefferson County/Westminster, Colorado, Cultural Resource Survey, 2008–2009* (Littleton, CO: Bunyak Research Associates, 2009), 23–24; and Linda Graybeal, "Semper: The People, the Place, the Preservation," *Historically Jeffco* no. 34 (2013): 2–6.

52. *Denver Post*, September 3, 1911, 9; and *Rocky Mountain News*, September 4, 1911, 4.

53. The Denver and Interurban (D&I) Railroad operated on a separate

electrified track that paralleled the Colorado and Southern from Globeville northwestward to Boulder. Between downtown Denver and Globeville D&I trains (typically just a single car) ran on the tracks of the Denver Tramway Company, which had an exclusive franchise for streetcar service within city limits. According to a July 1908 schedule, D&I cars departed downtown Denver every hour on the hour, arrived at Globeville in fifteen minutes, and arrived at Semper fourteen minutes later. William C. Jones and Noel T. Holley, *The Kite Route: Story of the Denver & Interurban Railroad* (Boulder, CO: Pruett Publishing Co., 1986), 19, 107.

54. Denver Tramway received five cents for the ride out to Globeville, and the D&I received twenty cents. Jones and Holley, *The Kite Route*, 18, 107.

CHAPTER 3

1. "Semper Garden Tracts," Colorado Bond and Realty Co., Inc., ca. 1908; Sheet number 29, Folder 5; Jefferson County Abstract Company Map Collection, 1867–1949; Series 143; Jefferson County Records Management & Archives, Golden, Colorado. Semper Garden Tracts was platted around 1896, according to Bunyak and others, *Westminster Selective Intensive Survey*, 32. Jefferson County archivist Stephanie Frasier noted that "the original of the [plat map] is long gone but I did find a hand-traced copy of the original showing 'Date Recorded August 28, 1908.'" Stephanie Frasier, email, August 21, 2019.

2. The locater inset was traced from US Geological Survey, *Arvada, Colorado* (quadrangle map), 1:31,680, 7.5-minute series, 1941.

3. The D&I track was electrified with poles and overhead wire, and its high-speed trains had a prearranged meeting at a designated passing siding between Denver and Boulder. The C&S dispatcher controlled both lines, but to avoid interference with schedules and operations, the D&I track was always on the east, north, or northeast side of the C&S track. The C&S track included an additional rail for the ore trains of the narrow-gauge Denver, Boulder, and Western Railroad, which had trackage rights. Jones and Holley, *The Kite Route*, 9–14, 28–29.

4. The first depot, shown on the 1896 subdivision map as just north of the pointed corner of lot 13, was probably closed after the C&S lowered the track to reduce the gradient as well as mollify any "undulation" (rise and fall) that might have interfered with the efficient operation of its steam locomotives. For discussion of betterment projects, see Mark Monmonier, *Connections and Content: Reflections on Networks and the History of Cartography* (Redlands, CA: Esri Press, 2019), 110–11; and William G. Raymond, *Elements of Railroad Engineering*, 3rd ed. (New York: John Wiley & Sons, 1917), 348–52.

5. Relocating the station was easier than building a bridge because a typical interurban station was a raised platform, with or without a roof. On June 1, 1908, local residents sent the county commissioners a petition protesting the railroad's "closing a portion of the public highway at and near Semper station . . . and shut[ting] off this crossing to the public causing travelers much danger, inconvenience, and loss of time." They called for the C&S to provide a "safe and suitable crossing over their track," but the railroad chose instead to move the station. See "Petition to the Honorable Body of County Commissioners of Jefferson County, Colorado, June 1, 1908," signed by nineteen residents of the Semper neighborhood. Provided by Archivist, Jefferson County, August 21, 2019.

6. A 1908 D&I employee timetable placed the interurban stop at the grade crossing closer to the lower right, within the dashed-line circle. Although the dashed circle might suggest four possible locations at the grade crossing, the platform was most certainly in one of the two corners northeast of the track.

7. Jessie E. Ferguson to John B. Plato, June 19, 1918, warranty deed, book 218, 85. Located less than a half mile from his original ten-acre property, it is bounded by the railroad on the southwest and a public highway on the northwest. It remained in his estate until nearly seven years after his death. Purchase price was "One Dollar and other valuable consideration," sometimes a strategy for not revealing the purchase price in a public document. The lot was also described in a warranty deed recorded when Plato's wife sold the property for $2,500 several years after his death. Christine G. Plato to Daniel J. Donmyer and Carolyn K. Donmyer, January 31, 1973, warranty deed, book 2480, 118. Copies of both deeds provided by the Archivist, Jefferson County, Colorado, July 24, 2019, also the source of the 1:6,000 9 × 9-inch air photo, frame number 152, dated May 14, 1956, in photo set 116-6.

8. In 1919 the Federal Railroad Administration assessed agricultural prospects in Colorado and offered a favorable assessment of irrigated agriculture in the Foothills Region, which includes Semper. See "Livestock Production Is Successful throughout the Region, Especially Dairying," US Railroad Administration, *Colorado: An Undeveloped Empire*, Agricultural Series no. 9 (Chicago: Poole Bros., 1919?), quotation on 24. For a more recent discussion of the challenges of dry farming, see William F. Schillinger, Robert I. Papendick, Stephen O. Guy, Paul E. Rasmussen, and Chris Van Kessel, "Dryland Cropping in the Western United States," in *Dryland Agriculture*, 2nd ed. Agronomy Monograph no. 23, 365–93 (Madison, WI: American Society of Agronomy; Crop Science Society of America; and Soil Science Society of America, 2006).

9. Bunyak and others, *Westminster Selective Intensive Survey*, 23–24; and Graybeal, "Semper," 2–6.

10. Jefferson County, Colorado, "Historic Land Records," book 175, 546, tract 4, Semper Garden Tracts, Benjamin Kaufman and Abraham Cohen to John B. Plato, January 21, 1911, warranty deed; and book 185, 365, tract 4, Semper Garden Tracts, John B. Plato to Benjamin Kaufman and Abraham Cohen, January 2, 1913, warranty deed.

11. Jefferson County, Colorado, "Historic Land Records," book 189, 269, tracts 9, 11, and 13, Semper Garden Tracts, W. R. Torbert to John B. Plato, March 1, 1916, deed of trust.

12. The original Semper Garden Tracts map shows a right of way running north–south between lots 1 and 2, 3 and 4, and so forth. It might have been the locus of a footpath or even an unimproved dirt road for wagons.

13. James Edward Le Rossignol, *Taxation in Colorado* (Denver: G. T. Bishop, 1902), 7, 13; "Instructions to County Assessors and Syllabus of New and Amended Laws Concerned with the Administration of the General Property Tax," Colorado Tax Commission, 1913; and G. S. Klemmedson and C. C. Gentry, *Outline of Colorado Tax Laws for Farmers and Ranchmen*, Bulletin 355 (Fort Collins: Colorado Agricultural College, Colorado Experiment Station, 1929), 8.

14. Assessment rolls were provided by the Jefferson County Archivist.

15. Clyde A. Hall and John B. Plato, "Reducing Barn Costs—Plans Used by Two Dairymen: One Has 30 Cows, the Other 11," *System on the Farm* 1, no. 5 (July 1917): 192–93; quotation on 193.

16. "Officers Elected: Jefferson County School Districts Select Secretaries and Fill Other Vacancies," *Colorado Transcript* (Golden, CO), May 11, 1916, 1.

17. *Denver Post*, May 24, 1914, 46, and August 2, 1914, 47.

18. J. B. Plato, "A Sensible Colorado Dairyman," *The Rural New-Yorker*, January 31, 1914, 162.

19. "Transfers of Guernsey Cattle from March 1 to March 14, 1918," *Guernsey Breeder's Journal* 13 (1918): 498–99; "Transfers of Guernsey Cattle from January 26 to February 10, 1920," *Guernsey Breeder's Journal* 17 (1920): A-5. Semi-monthly lists appear toward the end of a volume. The origin of "Featherton" is obscure but it provided a euphonious trademark. Plato might have had a prolonged impact on purebred bloodlines, albeit for Polled Herefords: on January 27, 1967, the *Xenia (OH) Daily Gazette* advertised a public sale of ninety-five cattle, fifteen calves, and ten bulls, "most sired by J. B. Plato Mischief 2nd, a 2500-lb. bull."

20. F. W. French, "Watch Locates Neighboring Farmers," *Illustrated World* 27 (April 1917): 247–48. The article incorrectly reports the township name as Bloomfield (with an L); it should be Broomfield (with an R).

21. *Illustrated World* reported his address as "Box 41, R.F.D. 1, Broomfield, Colorado," but his direct correspondence with the Patent Office shows his

address as "Featherton Farm, Route 1, Box 112, Broomfield, Colorado."

22. French, "Watch Locates Neighboring Farmers," 248.

23. Quotation from French, "Watch Locates Neighboring Farmers," 247.

24. John B. Plato to Commissioner of Patents, December 3, 1914. For further discussion of Plato's interaction with the Patent Office, see Monmonier, *Patents and Cartographic Inventions*, 31–36.

25. L. D. Underwood to John Byron Plato, January 16, 1915.

26. Rollandet to Commissioner of Patents, February 25, 1915.

27. Underwood to Rollandet, March 31, 1915.

28. Plato's original longhand application (in the National Archives' case file for patent 1,147,749) describes two drawings: the first was "a map subdivided according to my system" and the second was a "transparency carrying the necessary lines and numbers."

29. J. B. Plato, "Map or Chart," US Patent 1,147,749; filed December 7, 1914, and issued July 27, 1915.

30. Plato, "Map or Chart."

31. Plato, "Map or Chart."

32. *Clock System Rural Index, Ulysses Township* (Ithaca, NY: American Rural Index Corporation, 1919), quotation on 33 (inside back cover).

33. French, "Watch Locates Neighboring Farmers," 247–48, quotation on 248.

34. Stephen A. Kochersperger, Senior Research Analyst, Postal History, US Postal Service, emails, August 17, 2017, and August 28, 2019.

35. "Novel Scheme for Rural Box Numbers; Postal Department Is Considering Plan; Direction and Distance of the Box from the Postoffice Will Be Told," *Tacoma* (WA) *Daily Ledger*, November 21, 1915, 8.

36. "Government May Adopt Plato's Plan," *Colorado Transcript* (Golden, CO), November 25, 1915, 1.

37. "Finds Address by Looking at Watch," *Colorado Transcript* (Golden, CO), February 3, 1916, 1, and its continuation "Invents Plan for Mail Box Numbering," 8.

38. *Clock System Rural Index, Ulysses Township*, 33.

39. According to a 1912 timetable, Fort Collins was sixty-five miles north of Semper on the C&S Railway. Steam trains leaving Semper at 8:42 A.M. and 4:23 P.M. arrived in Fort Collins at 11:25 A.M. and 7:15 P.M. Return trips leaving at 9:05 A.M. and 2:40 P.M. arrived in Semper at 11:45 A.M. and 5:25 P.M. *Official Guide of the Railways, and Steam Navigation Lines of the United States, Porto Rico, Canada, Mexico and Cuba* 44, no 8 (January 1912): 608.

40. Population data from Colorado Department of Local Affairs, Historical Census Data—Counties and Municipalities, https://demography.dola.colorado

.gov/population/data/historical_census/. However impressive the population increase during the century's first decade, corresponding counts of 8,755 and 27,852 for 1920 indicate that growth had plateaued, at least for a while.

41. "Name or Number the Farm," *Daily Ardmoreite* (Ardmore, OK), September 26, 1916, 8. The article's Chicago dateline suggests it might have been distributed by a wire service.

42. "What Is Known as the 'Clock System'," *Topeka* (KS) *State Journal*, December 25, 1916, 5.

43. German-Russians began arriving in the 1890s. They had large families, and parents typically put their children to work at a young age. With a strong work ethic, the German-Russians assimilated rapidly and many bought farms. Mary Lyons-Barrett, "Child Labor in the Early Sugar Beet Industry in the Great Plains, 1890–1920," *Great Plains Quarterly* 25, no. 1 (2005): 29–38.

44. US Geological Survey, *Fort Collins, Colorado* (quadrangle map), 1:62,500, 15-minute series, 1908.

45. For instance, the 1908 USGS map shows a comparatively abrupt bend in sec. 4, T.8N, R.69W, after C&S tracks trending SW–NE enter the section on the west side and turn due north just south of the center of the section. Although the smoother, broader curve on Plato's map might suggest a C&S betterment project, the original, more abrupt curve was readily apparent five decades later on USGS, *Fort Collins, Colorado* (quadrangle map), 1:24,000, 7.5-minute series, 1960.

46. R. W. Gelder, *Map of the Irrigated Farms in Northern Colorado, 1915*, item CO00113, Historic Maps Collection, Fort Collins Museum of Discovery and the Poudre River Public Library District. Description and scans online at https://history.fcgov.com/collections/historic-maps. The Greeley Municipal Archive also has a copy.

47. For an alphabetical list of landowners compiled from the irrigation map, see Marilyn Reed and Jacquelyn Glavinick, Index to Irrigated Farm Owners of Northern Colorado, 1915, Fort Collins History Collection, https://fchc.contentdm.oclc.org/digital/collection/hm/id/737/rec/18.

48. A conscientious search by a staff member at the Library's Geography and Map Division turned up only one of Plato's maps (for Erie County, Pennsylvania) deposited with a copyright application.

CHAPTER 4

1. William Wyckoff, "Cartography and Capitalism: George Clason and the Mapping of Western American Development, 1903–1931," *Journal of Historical Geography* 52 (2016): 48–60.

2. Though not common, maps submitted for copyright registration occasionally have been cut to focus the registration on a specific part of the artifact, or perhaps to exclude an element for which a copyright was not claimed. John W. Hessler, Curator, Jay I. Kislak Collection of the Archaeology and History of the Early Americas, Geography and Map Division, US Library of Congress, email, August 28, 2019.

3. Thanks to Geography Department colleague Jacob Bendix, the first friend to respond after I posted the image on Facebook with my own admittedly illogical interpretation, "LOCKED BY."

4. *Hollister's Rural Index* (Chicago: H. L. Hollister, 1917), quotation on 1 (interior right-hand page).

5. Hiram T. French, *History of Idaho: A Narrative Account of Its Historical Progress, Its People and Its Principal Interests* (Chicago: Lewis Publishing Co., 1914), 3: 1307–8; photo on unnumbered page facing 1307; quotation on 1308.

6. *Putting the Farmer on the Map: Hollister's Rural Index* (Chicago: H. L. Hollister, 1917).

7. Hollister made no attempt to register a copyright, which would have affirmed the publication date. However, "JAN 18 1918," stamped on page 3 of *Putting the Farmer on the Map*, is several months later than the "AUG 16 1917" due-date stamped on a card in the back of the folio-size *Hollister's Rural Index*. Also noteworthy, the *Catalog of Copyright Entries* reported that the forty-four-page version held by the USDA had been published on June 6, 1917, nearly two months after Hollister's release of a thirty-eight-page "Preliminary Prospectus" on April 10. Library of Congress, Copyright Office, *Catalog of Copyright Entries*, Part 1: Books, Group 2, n.s., vol. 14, nos. 5 and 7 (Washington, DC: Government Printing Office, 1917), 490 and 747.

8. *Putting the Farmer on the Map*, 3.

9. *Putting the Farmer on the Map*, 6.

10. *Putting the Farmer on the Map*, 6–7.

11. *Putting the Farmer on the Map*, 7.

12. *Putting the Farmer on the Map*, 7.

13. *Putting the Farmer on the Map*, 7.

14. *Clock System Rural Index, Ulysses Township*, 33 (inside back cover).

15. "Facts Puncture $5,000,000 Dream; Big Men Involved: A. B. Hulit's Scheme for Chicago Agricultural Exposition Shattered," *Chicago Tribune*, August 12, 1912, 1, 4.

16. "Proposed Nation-Wide Survey of Agricultural Interests," *Wisconsin State Journal*, April 18, 1917, 7.

17. "To Index Farm Homes," Stevens Point (WI) *Gazette*, May 9, 1917, 7.

18. Charles Josiah Galpin, *My Drift into Rural Sociology* (Baton Rouge: Louisiana State University Press, 1938), esp. 3–35; J. H. Kolb, "Dr. Galpin at Wisconsin," *Rural Sociology* 13 (1948): 130–45; Henry C. Taylor, "The Development of Country Life Studies at the University of Wisconsin," *Rural Sociology* 6 (1941): 195–202.

19. Charles J. Galpin, *Rural Life* (New York: Century, 1918), quotation on 341.

20. Galpin, *Rural Life*, 341–42.

21. Galpin explored two complementary strategies—interviewing "physicians, livery men, real estate men, ministers, bankers, and the like" within the center and interviewing individual farmers—and found their results were essentially similar. Galpin, *Rural Life*, 338–40, quotation on 338.

22. Galpin, *Rural Life*, 341.

23. Galpin, *Rural Life*, 342.

24. Galpin, *Rural Life*, 342.

25. W. A. Anderson, "Dwight Sanderson, Rural Social Builder," *Rural Sociology* 11, no. 1 (1946): 7–14; and Carl C. Taylor, "Dwight Sanderson—Social Scientist," *Rural Sociology* 11, no. 1 (1946): 14–23.

26. Dwight Sanderson, "Locating the Rural Community," *Cornell Reading Course for the Farm*, lesson 158 (June 1920): 415–36, esp. 429–33, quotation on 429.

27. Sanderson, "Locating the Rural Community," 433.

28. Sanderson, "Locating the Rural Community," 422–25.

29. It would be more accurate to say that I redrafted Plato's map to ensure equal angles between sectors and equal spacing between circles. Although his clock grids look right at a normal viewing distance, close inspection revealed small irregularities, probably because he was using mechanical drawing instruments less precisely than I was using Adobe Illustrator. Sanderson's drafting was no less precise, and his map misrepresented the town boundary by squaring off the pointed extension in the northwest.

30. *Clock System Rural Index, Ulysses Township*, 33.

CHAPTER 5

1. Although trim size varies noticeably, the pages of Plato's directories are noticeably taller than the 11 inches of letter-size paper. According to the *Catalog of Copyright Entries*, the Ulysses directory had a quarto format ("4to"), which typically means 9½ × 12 inches, but these dimensions are approximate. In general, the longer dimension is slightly less than 12 inches and the shorter dimension is closer to 8½ inches; 9 × 12 inches seems a workable compromise.

2. M. P. Goodhue, compiler, *Norton and Goodhue's Ithaca City Directory for 1917–18* (Ithaca, NY: Norton Printing Co., 1917), map on inside back cover.

3. Index Map Company, *Clock System Map of Tompkins County, New York* (Ithaca, NY, 1924).

4. *Clock System Rural Index, Ulysses Township*, inside back cover.

5. The origin of this saying is obscure. Though clearly relevant to commercial television in the pre-cable era, and to radio even earlier, it is often applied to services like Facebook and Google. For discussion without an exact citation, see how John Lancaster explores the theme in "You Are the Product," *London Review of Books* 39, no. 16 (August 17, 2017): 3–10.

6. Advertisement for Universal Sales Co., 110 West Green Street, Ithaca, *Clock System Rural Index, Ulysses Township*, 22.

7. *Clock System Rural Index, Ulysses Township*, inside front cover. Also see Andrew Denny Rodgers III, "Bailey, Liberty Hyde, Jr." *Dictionary of Scientific Biography*, 1:395–97.

8. *Clock System Rural Index, Ulysses Township*, 6.

9. *Clock System Rural Index, Ulysses Township*, 10.

10. *Clock System Rural Index, Ulysses Township*, 4.

11. For Bailey's recollection, see *Clock System Rural Index, Ulysses Township*, inside front cover. For the byline, see J. B. Plato, "Numbering Farm Houses," *Cornell Countryman* 15, no. 7 (April 1918): 401–2, 422.

12. John Byron Plato, United States Selective Service System Draft Registration Cards, World War I, accessed through Ancestry Library Edition. Plato and his mother were fortunate to have survived the influenza epidemic blamed for 140 deaths in Tompkins County in the last four months of 1918; see Charley Githler, "Outbreak: Ithaca and the 1918 Flu Epidemic," Ithaca.com, January 26, 2018; https://www.ithaca.com/news/outbreak-ithaca-and-the-flu-epidemic/article_854e6bfa-0207-11e8-b816-338345014cef.html (accessed March 21, 2020).

13. With operations at three sites in Ithaca, the firm was a major local employer. For a contemporary description of the Thomas-Morse Aircraft Corporation, see Fay L. Faurote, ed., *Aircraft Yearbook, 1919* (New York: Manufacturers Aircraft Association, 1919), 240–69.

14. Plato still had one foot on the farm, at least through 1919, when he failed to pay his property tax, defaulted on his mortgage, and lost his lots—that is, all but the 0.4-acre triangular lot in sect. 24 (fig. 12) that he had bought on June 19, 1918 (Book 219, p. 85, Jefferson County Colorado, Archives). Indeed, intermittent short-term tax delinquency was a pattern apparent throughout his residence at Semper. In 1919 he defaulted on lots 1 and 3 (Book 210, pp. 498–500); they were last listed in the tax rolls for 1919. He also lost lots 9, 11, and 13, which went back to Henry & Minnie Weinsheimer (Book 212, p. 536, January 31, 1920). The

Weinsheimers also redeemed lots 1 and 3 on January 31, 1920 (Assessors Index Book, Block: Semper Garden Tracts, Plots 1 to 13).

15. Plato's draft card reported Helen's residence as 616 Franklin Street, Geneva, Kane County, Illinois. According to the 1920 Census, this was the home of her sister-in-law, Helen P. Wilbur. Enumerator's schedules for the 1920 Census accessed at Ancestry Library Edition.

16. State certification was the culmination of a process initiated at the local county clerk's office. The firm's Certificate of Incorporation, recorded February 26, 1919, includes a mission statement and the names, addresses, and numbers of shares of the original eight shareholders. Plato held fifteen of the one hundred shares. Tompkins County Clerk backfile nos. 037300 and 037578. An amendment recorded on August 27, 1919, increased the capitalization (total par value of shares that could be issued) to $10,000. Backfile nos. 037557 and 037283. A further amendment recorded on January 11, 1922, increased the firm's capitalization to $30,000. Backfile no. 037360.

17. "Index Corporation Chartered by State," *Ithaca Journal*, March 1, 1919, 6. An eighth "incorporator," Sydney L. Tuttle (an official of the Ithaca Condensed Milk and Cream Co.), was not a director.

18. For discussion of the corporation's Cornell ties, see "The Rural Index," *Cornell Alumni News* 22, no. 17 (January 22, 1920): 194–95.

19. "Locates Farm Homes Through Clock System," *Ithaca Journal*, January 18, 1919, 7; "County Farms Identified by Clock System," *Ithaca Journal*, November 6, 1920, 6.

20. The literature on business models is voluminous, complex, and mostly focused on start-up software, internet, and consumer electronics firms. My discussion is based on elements of the Business Model Canvas devised by Strategyzer AG, a Swiss firm; see https://strategyzer.com/canvas/business-model-canvas (accessed September 18, 2019). For a critical overview of business models, see Carlos M. DaSilva and Peter Trkman, "Business Model: What It Is and What It Is Not," *Long Range Planning* 47 (2014): 379–89; and Lars Schweizer, "Concept and Evolution of Business Models," *Journal of General Management* 31 (2005): 37–56.

21. *Clock System Rural Index, Ulysses Township*, 6.

22. *Clock System Rural Index, Ulysses Township*, 33 (inside back cover).

23. *Clock System Rural Index, Ulysses Township*, 10.

24. *Clock System Rural Index, Ulysses Township*, 6.

25. I know Plato used a pantograph because he said so, in a 1927 *Ithaca Journal* profile of his business. "Know Ithaca: The Index Map Co., Inc.," *Ithaca Journal*, October 12, 1927, 13.

26. *Clock System Rural Index, Ulysses Township*, 6.

27. *Clock System Rural Index, Ulysses Township*, 6.

28. *Clock System Rural Index, Ulysses Township*, 8.

29. *Clock System Rural Index, Ulysses Township*, 10.

30. Statement of George R. Fitts, President, Tompkins County Farm Bureau, *Clock System Rural Index, Ulysses Township*, 1.

31. *Clock System Rural Index, Ulysses Township*, 8.

32. "Locates Farm Homes Through 'Clock' System," *Ithaca Journal*, January 18, 1918, 7.

33. "Ithacan's Invention to Put Farmer on the Map Finds Place in Digest," *Ithaca Journal*, February 20, 1920, 7; and "Putting the Farmer on the Map," *Literary Digest* 64, no. 8 (February 21, 1920): 28–29.

34. "Rural Index Work Progressing Steadily," *Ithaca Journal*, May 25, 1921, 5.

35. "Farmers Win $200.00 in Premiums at Fair" (advertisement), *Ithaca Journal*, September 23, 1919, 5.

36. "Today's Fair Entirely Unlike Early Exhibits," *Ithaca Journal*, August 25, 1920, 5.

37. "15,000 Admissions New Record for Fair Attendance; Continue Pageant Friday," *Ithaca Journal*, September 1, 1921, 5.

38. "Social and Personal," *Ithaca Journal*, May 8, 1920, 3; October 21, 1920, 3.

39. "Social and Personal," *Ithaca Journal*, August 30, 1920, 3.

40. "Scout Notes," *Ithaca Journal*, November 30, 1921, 3; "Scout Troop 15 Installed at Public Meeting," *Ithaca Journal*, December 9, 1921, 7.

41. "22 Members in New Ad Club Organized Here," *Ithaca Journal*, May 10, 1924, 7; "Ithaca 'Ad' Club All Ready for First Birthday Party," *Ithaca Journal*, May 19, 1925, 12.

42. "Skating Contests for Children To Be Held Saturday," *Ithaca Journal*, February 24, 1926, 5; "Kiddies Enjoy Skating Events on Beebe Lake," *Ithaca Journal*, March 1, 1926, 5.

43. Samuel T. Williamson, *Frank Gannett: A Biography* (New York: Duell, Sloan and Pearce, 1940), esp. 72–73.

44. J. B. Plato, "Numbering Farm Houses," *Cornell Countryman* 15, no. 7 (April 1918): 401–2, 422.

45. "The Rural Index," *Cornell Alumni News* 22, no. 17 (January 22, 1920): 194–95.

46. *Cornell Alumni News* 24, no. 14 (January 12, 1922): 169.

47. "Rural Index Co. Will Push Clock System; Frank E. Gannett Elected President of Corp. at Annual Meeting," *Ithaca Journal*, January 7, 1922, 5.

48. This location, at the southeast corner of N. Tioga and E. Seneca Streets,

was also known as the Library Building because it hosted the city's public library on the second floor. By the mid-1930s, the bank had relocated to a new First National Bank Building, on the northeast corner of E. State and N. Tioga Streets.

49. "Official Numbers for Every Farm House in Tompkins County," *Ithaca Journal*, July 21, 1920, 7.

50. For a history of Ithaca's railway connections, see Hardy Campbell Lee, Winton Rossiter, and John Marcham, *A History of Railroads in Tompkins County*, 3rd ed. (Ithaca, NY: DeWitt Historical Society of Tompkins County, 2008).

51. Qualitative judgments about relative neighborhood quality are based on a visual canvass, albeit nearly a century later, as well as the reported occupations of Plato's neighbors. After 1922, fewer households in his vicinity had live-in maids and more residents who boarded ("b," indicating that they also took meals at the house) or merely rented a room ("r"). City directories were published biennially through 1919–20 by Norton & Goodhue, and annually for 1921 and 1922, and then for October 1923, January 1925, May 1927, and annually starting January 1929, by H. A. Manning.

52. The location was described as "Adams c Dey," meaning on the corner of Adams and Dey Streets. *Ithaca (New York) Directory for the Year Beginning October 1923* (Schenectady, NY: H. A. Manning Co., 1923), 444.

53. The Index Map Company, Inc. was a new company with an authorized capitalization of $20,000, with one thousand common (voting) shares with a par value of $10 per share and one thousand preferred shares with a par value of $10 per share. Earnings would first be paid to holders of preferred shares at a cumulative rate of 8 percent per annum, and any shortfall would be paid out of earnings during subsequent years. Only then would earnings be paid to holders of common shares. Holders of preferred shares were allowed to vote only on matters related to changing the amount of preferred stock and the rights of preferred shareholders, all of whom had to approve of any changes. Initially, the firm had three directors: John B. Plato (ninety-eight shares), Helen L. Plato (one share), and local attorney E. Morgan St. John (one share). The certificate of incorporation was signed on March 26, 1924, and recorded on April 19, 1924. Tompkins County Clerk backfile no. 037450.

54. Classified advertising, under Financial / Investments, *Ithaca Journal*, May 11, 1926, 13.

55. "State Issues Charter to Index Map Company Advertising Concern," *Ithaca Journal*, April 4, 1924, 7. Advertising was the dominant theme of the new company's statement of purpose, which began "To purchase, sell and deal generally in all patents and copyrights connected with the establishment and

management of the rural and urban directory and advertising business" and mentioned "maps" only fifty-seven words later—and only once unless one counts "road maps and guides" in the litany of products "necessary or useful in carrying on the directory and advertising business." Tompkins County Clerk backfile no. 037450.

56. "Car Lands in Owego Creek; Ithacan Hurt," *Ithaca Journal*, September 1926, 7; "2 Industrial Tours Slated Thursday," *Ithaca Journal*, September 28, 1927, 5.

57. Quotations in this and the following three paragraphs are from "Know Ithaca: The Index Map Co., Inc." Among other leads, this story was the first concrete indication of Plato's learning about farming in Cornell's Winter Course.

58. "Know Ithaca: Ithaca Engraving Co., Inc.," *Ithaca Journal*, August 31, 1927, 6.

59. For a more or less contemporaneous description, see Erwin Raisz, *General Cartography* (New York: McGraw-Hill Book Co., 1938), 172–91, esp. 181–82.

60. "Know Ithaca: The Index Map Co., Inc."

61. "Know Ithaca: The Index Map Co., Inc."

62. Under the Copyright Act of 1909, which was in force at the time, unless an accident could be confirmed, a work published without a copyright notice was automatically in the public domain. Susan M. Bielstein, *Permissions, A Survival Guide: Blunt Talk about Art as Intellectual Property* (Chicago: University of Chicago Press, 2006), 20–21; *The Chicago Manual of Style*, 17th ed. (Chicago: University of Chicago Press, 2006), 187. Moreover, a copyright not filed "promptly" would also be voided. Although the law did not specifically define "prompt," mailing the application and deposit copies "immediately after publication" was a prudent practice. Even so, the astute publisher retained a postal receipt or proof of mailing should the package be destroyed in transit. For a contemporary reference, see Richard Rogers Bowker, *Copyright: Its History and Its Law* (Boston: Houghton Mifflin Co., 1912), 142.

63. *Clock System Map of Edinboro Community, Erie County, Penna: Territory Covered by Edinboro Vocational School* (Ithaca, NY: American Rural Index Corporation, 1921). The map's entry in the *Catalog of Copyright Entries* reported publication on September 15, 1921, thirteen months before the firm published a map and sixty-page directory covering the entire county. Library of Congress, Copyright Office, *Catalog of Copyright Entries*, Part 1: Books, Group 2, n.s., vol. 18, no. 11 (Washington, DC: Government Printing Office, 1921), 1429; and vol. 19, no. 12, 1726. The enumerator's manuscript for the 1920 Census, accessed through Ancestry Library Edition, reported Russell D. McCommons, age fifteen in early 1920, as an Edinboro resident living with four siblings in the home of

his mother, a forty-two-year-old dressmaker. The enumerator's list for the 1940 Census identified him as a thirty-five-year-old married college teacher living in Nashville, Tennessee.

64. A note on the right below the frame indicates use of a non-local printer: "Reproduced and Printed by National Process Company, Inc., New York, U. S. A."

65. *"Clock System" Map of Cayuga County, New York* (Ithaca, NY: Index Map Company, 1925).

66. *"Clock System" Map of Onondaga County, New York*, series no. 16, ca. 1:72,000 (Ithaca, NY: Index Map Co., 1927).

67. Improved *"Clock System" Map of Oneida County, New York* (Ithaca, NY: Index Map Co., 1928).

68. At the lower left, stamped in blue ink, is the label "Gift: Chas. B. Peterson III." For the online catalog entry, see https://lccn.loc.gov/93680151.

69. I caution my students to always use graphic scales on detailed maps of small-to-medium size areas and to never use ratio scales because the map might be enlarged or reduced.

70. Malcolm Lloyd, who wrote a practical guide for engineers and students, considered Wrico pen-and-stencil sets "expensive, $27 for the set" and "not time savers as claimed." See Malcolm Lloyd, *A Practical Treatise on Mapping and Lettering* (Philadelphia: P. Blakiston's Son & Co., 1930), 5. For a historical overview, see Karen Severud Cook, "Labeling of Maps: Labeling Techniques," *HC6*, 738–44; Hans-Uli Feldmann, "Drafting of Maps: Drawing Instruments," *HC6*, 326–29; and Hans-Uli Feldmann, "Drafting of Maps: Pen-and-Ink Drawing," *HC6*, 331–34.

71. Among the many instances of this poem that could be cited is the three-column-wide ad, "The Tale of the 'Pig Tail Ad,'" *Ithaca Journal*, June 30, 1925, 13.

72. "The Tale of the 'Pig Tail Ad,'" *Ithaca Journal*, June 30, 1925, 13.

73. *"Clock System" Rural Index of Cayuga County, New York* (Ithaca, NY: Index Map Co., 1926), quotation on cover.

74. For examples see *1926–27 Special Classified and Graded List of All Farms of Broome County, New York, for Merchants and Manufacturers* (Ithaca, NY: Index Map Co., 1926) and *1928–29 Classified and Graded Directory of All Farms and Rural Homes in Cayuga Co., N. Y.* (Ithaca, NY: Index Map Co., 1928). Other counties with a similar list include Chenango (1927–28), Genesee (1929), Oneida (1928–29), and Onondaga (1928–29). Quotations from cover and first page of the guide.

75. "Know Ithaca: The Index Map Co., Inc."

76. Classified "Help Wanted—Female" (ad), *Ithaca Journal*, March 15, 1922, 8.

77. The 1922 city directory identified Baker as a "com trav" (commercial traveler) as well as the proprietor of Cayuga Farms, which sold dairy products.

Manning's Ithaca (New York) Directory, 1922 (Schenectady, NY: H. A. Manning, 1922), 44.

78. The 1929 city directory listed Tyler's employment as "sec Index Map Co and prop The Lion Realty Co." *Manning's Ithaca (New York) Directory for the Year Beginning January 1929* (Schenectady, NY: H. A. Manning, 1929), 387. His firm shared room 202, on the floor below Plato, with several other tenants.

79. *Manning's Ithaca (New York) Directory for the Year Beginning May 1927* (Schenectady, NY: H. A. Manning, 1927), 427–28.

80. "Mrs. Helen Larrabee Plato," *Ithaca Journal*, 5 March 1931, 5.

81. *Manning's Ithaca (New York) Directory for the Year Beginning January 1931* (Schenectady, NY: H. A. Manning, 1931), 381.

82. *Manning's Ithaca (New York) Directory, for the Year Beginning January 1932* (Schenectady, NY: H. A. Manning, 1932), 374 and 563.

83. *Manning's Ithaca (New York) Directory for the Year Beginning January 1933* (Schenectady, NY: H. A. Manning, 1933), 242 and 373.

CHAPTER 6

1. "Farms Are Given Street Numbers; Hancock County Directory to Describe Location of Farms by Number," *Findlay* (Ohio) *Morning Republican*, July 16, 1930, 7. Plato's Ohio phase was more difficult to reconstruct than his earlier years in Colorado and New York. I had to rely largely on intermittent newspaper stories like this one, which provided the first and strongest clues that he had not fallen ill or was wandering aimlessly between leaving Ithaca and landing in Washington in the middle of the Great Depression.

2. With slightly more than 2,900 farms in 1930 and an average farm size of 107 acres, Hancock County was part of one of the state's two leading areas for beef and corn. J. I. Falconer, *Twenty Years of Ohio Agriculture, 1910–1930*, Bulletin 526 (Wooster, OH: Ohio Agricultural Experiment Station, 1933), 58 and 82.

3. "Numbers Given to Farm Homes," *Findlay* (Ohio) *Morning Republican*, September 9, 1930, 3.

4. "New Rural Index and County Farm Map" (ad), *Findlay* (Ohio) *Morning Republican*, September 12, 1930, 2.

5. The address might have been changed when he paid the tax on his 1930 assessment in October 1931. Ronda Frazier, Jefferson County (Colorado) Archives, email, August 2, 2019.

6. "Farm Directory Completed Here," *Findlay* (Ohio) *Morning Republican*, January 30, 1931, 10.

7. I queried Newspapers.com and NewspaperArchive.com, as well as WorldCat.org and the *Catalog of Copyright Entries*. Digitized newspapers are

apparently lacking for 1930–1935 for Ashland and Wayne Counties, as well as many other Ohio counties. Plato apparently never registered copyrights for his Ohio maps. Mostly fruitless email queries were sent to the county public library and county historical society in each of the four counties, and also to the state archivist, the state historical society, and the map library at the Ohio State University. Although I confirmed directories for four counties (Hancock, Licking, Wayne, and Ashland), the line "Ohio Series No. 5" on the cover of the Ashland directory suggests there might have been at least one other. I doubt strongly that additional examples would provide new information about the content of Plato's Ohio maps or how he made them. Indeed, an email plea to county libraries and historical societies within a plausible twenty-eight-county search area in northern Ohio was fruitless.

8. "Farm Index Map Shows Each Licking County Rural Home," *Newark* (Ohio) *Advocate*, October 14, 1931, 2.

9. Ohio's state archives, now known as the Ohio History Connection, supplied a copy of "Articles of Incorporation of Farm Index Service, Incorporated," October 23, 1931, corporation number 148516, Ohio State Archives Series 3080, vol. 400, 206–7. I discovered the date of the business charter in "Ohio Charters," *Cincinnati Enquirer*, October 26, 1931, 14.

10. Newspaper databases provided additional information about Plato's two partners: (list of activities at Dalton High School), *Orville* (Ohio) *Courier Crescent*, November 3, 1930, 5; "Lyman R. Critchfield, Jr.," (ad by Democratic Executive Committee; Critchfield was a candidate for county prosecuting attorney), *Orville* (Ohio) *Courier Crescent*, November 3, 1930, 3.

11. For further information, see the Wikipedia articles "National Union Catalog" and "OCLC."

12. *Rural Index and Buying Guide for Wayne County, Ohio, for 1931 and 1932* (Johnstown, OH: Farm Index Service, Inc., 1931), 14.

13. "Asks How to Pay," *Newark* (Ohio) *Advocate*, November 9, 1931, 2.

14. "Claims Judgment," *Newark* (Ohio) *Advocate*, December 4, 1931, 19.

15. "Claims Illegal Levy," *Newark* (Ohio) *Advocate*, December 16, 1931, 2.

16. It was not clear how Bowers was able to sell a vehicle on which Chippewa had a lien. Current ownership titling practices would probably have prevented the sale.

17. "Decree for Defendant," *Newark* (Ohio) *Advocate*, December 22, 1931, 9.

18. "Notice of First Meeting of Creditors," *Newark* (Ohio) *Advocate*, February 12, 1932, 14.

19. Dated November 15, 1932 and stamped in the upper right corner of the first page of the Articles of Incorporation, the concise five-line cancelation notice

referenced article 5509 of the Ohio G. C. (General Code).

20. "Sustained Demurrer," *Newark* (Ohio) *Advocate*, March 25, 1933, 2.

21. "Denies Allegation," *Newark* (Ohio) *Advocate*, May 12, 1933, 8.

22. "Files Demurrer," *Newark* (Ohio) *Advocate*, April 22, 1933, 2.

23. "Decree for Defendant," *Newark* (Ohio) *Advocate*, May 16, 1933, 8.

CHAPTER 7

1. "Veterans of Spanish War Hear Seven Addresses," *Washington Evening Star*, March 25, 1934, 49.

2. *Boyd's District of Columbia Directory, 1935* (Washington, DC: R. L. Polk & Co., 1935), 1569.

3. After the Supreme Court declared a tax on producers unconstitutional, the AAA invoked land conservation as the pretext for paying farmers to take land out of production. Because production controls depended on reliable estimates of land in production, the AAA devised innovative strategies for using aerial photography to map and measure farmland. See Mark Monmonier, "Aerial Photography at the Agricultural Adjustment Administration: Acreage Controls, Conservation Benefits, and Overhead Surveillance in the 1930s," *Photogrammetric Engineering and Remote Sensing* 76 (2002): 1257–61; and Mark Monmonier, "Agricultural Adjustment Administration (U.S.)," *HC6*, 30–31.

4. *Boyd's District of Columbia Directory, 1936* (Washington, DC: R. L. Polk & Co., 1936), 1705.

5. Personal Data Memorandum completed by John B. Plato, November 27, 1934, Agricultural Adjustment Administration, USDA. In November 2017, the National Personnel Records Center, at the National Archives facility in St. Louis, provided copies of Plato's Civil Service records for his employment at the Department of Agriculture (including the AAA) and the Department of Commerce (which included the Census Bureau).

6. Carl C. Taylor, "Charles J. Galpin (1864–1947)," *American Sociological Review* 13 (1948): 104.

7. Personal History Statement, completed by John Byron Plato, for the Census Bureau, August 1, 1934. US National Archives, St. Louis.

8. "Memorandum for the Director," from Stuart A. Rice, Assistant Director, Bureau of the Census, Department of Commerce, July 30, 1934. US National Archives, St. Louis.

9. Employment status, "Plato, John B. (2/1/37)," Bureau of the Census, Department of Commerce. US National Archives, St. Louis.

10. Personal History Statement, completed by John Byron Plato, for the Census Bureau, August 1, 1934. US National Archives, St. Louis.

11. John B. Plato, Personal Data Memorandum, Resettlement Administration, July 16, 1935. US National Archives, St. Louis.

12. John B. Plato, Personal History Statement, Resettlement Administration, June 3, 1935. US National Archives, St. Louis.

13. John B. Plato, Statement for Personnel Record, Agricultural Adjustment Administration, July 31, 1934. US National Archives, St. Louis.

14. John B. Plato, Personal History, Bureau of Agricultural Economics, USDA, June 13, 1934. US National Archives, St. Louis.

15. Eric England, personnel recommendation to appoint John B. Plato as Clerk, CAF-4, in the Bureau of Agricultural Economics, USDA, May 2, 1934. US National Archives, St. Louis.

16. Organized by five-year time intervals, the eBook *Value of a Dollar* lists average salaries for numerous representative occupations. None of the twenty-six categories listed in "Standard Jobs 1935–39" has an average greater than $2,600, but the data lack estimates for professional and managerial work. Scott Derks, *The Value of a Dollar: Prices and Incomes in the United States, 1860–2014*, 5th ed. (Amenia, NY: Grey House Publishing, 2014).

17. See, for example, "Statement for Personnel Record: John Byron Plato," May 16, 1934, Bureau of Agricultural Economics, USDA. US National Archives, St. Louis.

18. "Memorandum to the Administrator," AAA, Vernon E. Bundy, Acting Director, Division of Information, July 27, 1934. US National Archives, St. Louis.

19. "Recommendation to the Secretary: Temporary Appointment of John B. Plato," Agricultural Adjustment Administration, USDA, July 28, 1934. US National Archives, St. Louis.

20. "Fact as Disclosed by Investigator's Report in the case of John Plato, applicant for [the] position assistant economist," AAA, USDA, August 14, 1935. US National Archives, St. Louis.

21. "Inter-Office Communication, from L. C. Gray, Assistant Administrator, to Mr. W. B. Stephens, Director, Personnel Division," Resettlement Administration, USDA, March 16, 1936. US National Archives, St. Louis.

22. "Recommendation to the Secretary: Temporary Appointment of John B. Plato," Agricultural Adjustment Administration, USDA, November 6, 1936. US National Archives, St. Louis.

23. Monmonier, "Aerial Photography at the Agricultural Adjustment Administration."

24. "Lincoln Parish Map Plan to Revolutionize Rural Route Systems of the Nation," *Ruston* (Louisiana) *Daily Leader*, February 3, 1937, 1, 4.

25. For discussion of stereo mats and casting boxes, see Mark Monmonier,

Maps with the News: The Development of American Journalistic Cartography (Chicago: University of Chicago Press, 1989), 79–82.

26. The newspapers were published in Corpus Christi, Texas; Davenport, Iowa; Hamilton, Ohio; and Washington Court House, Ohio. Because the databases I used were incomplete, there are probably others.

27. K. F. Hewins, "Government Begins Huge Task of Numbering Isolated Farms," *Quad-City Times* (Davenport, Iowa), February 12, 1937, 14.

28. See, for example, "Farmers Will Get Directory: Agriculture Department Is Making Experiment in Louisiana," *Charlotte Observer*, February 28, 1937, 57.

29. "Department Is Making Experiment in Louisiana," *Charlotte Observer*, February 28, 1937, 57.

30. The notice was pointedly explicit: "Mr. Plato's services will not be required after 11:45 A. M., October 13, 1937, and it is therefore recommended that his appointment be terminated effective at that time." "Termination of Temporary Appointment of John B. Plato, Section 2, Public No, 479, 74th Congress," Agricultural Adjustment Administration, USDA, September 21, 1937. US National Archives, St. Louis.

CHAPTER 8

1. "Memorandum for Mr. B. Z. Kile," December 12, 1934, J. B. Plato papers in the National Archives related to his USDA employment. US National Archives, St. Louis.

2. The lower left map relates Plato's immediate neighborhood on Berry Lane to Forestville, an unincorporated place more than a mile to the south along Forestville-Ritchie Road. (Though tree cover is not continuous, Forestville is an apt name for the area.) I redrafted this map from a 1944 USGS topographic map, which also supplied the much-enlarged image excerpt at the lower right: a more exact replica of the nearly dozen houses along the west side of the "unimproved road" represented by the double-dash line. This enlargement shows the horizontal label that the USGS used to designate Berry Lane as a neighborhood as well as a road; street names are customarily aligned with the feature. In both lower panels a light-gray tint mimics the USGS's light green symbol for woodland. The census taker's manuscript for the 1940 Census puts the Plato residence at the northern end of the street with the words "Here begins Berry Lane." 1940 Federal Census, Ancestry Library Edition.

3. Church Registers. Presbyterian Historical Society, Philadelphia, Pennsylvania, Ancestry Library Edition.

4. Christina Greiling, Declaration of Intent [to become a US citizen], US Immigration and Naturalization Service, no. 446603, November 16, 1939, New

York, State and Federal Naturalization Records, 1794–1943, Ancestry Library Edition.

5. According to the Social Security Death Index, Christine Plato had received a Social Security number (213-46-8697) in 1962. On December 29, 2014, I filed an electronic Freedom of Information Act (FOIA) request with the Social Security Administration, to obtain a copy of her original application, which would have included her place of birth. After receiving a "Numident Printout" that lacked key items, I followed up with a more detailed written letter on January 27, 2015, but my appeal led nowhere. No longer relevant to a plausible benefits claim, the information was simply not available.

6. Estate of John B. Plato, Prince George's County, Maryland, Register of Wills, estate no. RE-17,023, filed July 14, 1966, liber 1, pp. 224–25 (copies provided August 15, 2019).

7. A 1988 *Washington Post* story about the importance of volunteer visitors to nursing homes included a short interview with Christine, who was a resident of the Wilson Health Care Center, and apparently no longer living in her own home. Elisabeth McAllister, "Volunteers Become Tried and True Friends," *Washington Post*, November 10, 1988, MD26.

8. Estate of Christine G. Plato, Prince George's County, Maryland, Register of Wills, estate no. RE-38,479, filed April 16, 1991, esp. liber 139, pp. 177–78 (copies provided August 15, 2019).

9. See, for example, "Residential Zoning," *Washington Evening Star*, April 18, 1949, 26; and "Field Trip Planned for the Hilltoppers," *Washington Post and Times Herald*, May 5, 1957, E8.

10. "Japanese Iris Gay in 70-Acre Back Yard Site," *Washington Evening Star*, June 22, 1958, 26.

11. "Farmer, 84, Assails Bid for Apartment Zone," *Washington Post and Times Herald*, February 18, 1961, C2.

12. "Children Confer Title, 'Uncle John,' on Man Who Provided Forestville Camping Ground," *Washington Post*, July 16, 1947, 2B.

13. "Women's Clubs: Girl Scout Camp Activities Provided for More Than 1,200 This Summer," *Washington Evening Star*, August 9, 1951, B-6.

14. "Camping for a Day," *Washington Post*, May 11, 1952, 10S.

15. "Uncle John Plato, Benefactor of Girl Scouts in Forestville," *Washington Post*, July 1, 1966, B8.

16. "Maryland News in Brief," *Cumberland* (Maryland) *News*, April 10, 1967, 5.

17. "Camp Plato Place" is a brief note in the alphabetical list online at Girl Scout Council of the Nation's Capital, History and Archives Committee, https://archivesprojects.homestead.com/Camphistory.html. Also see the webpage

"Camp Plato Place," which has site maps at two different scales. https://archivesprojects.homestead.com/ArchivesCampPages/CampPlatoPlace.html. Also see Ann E. Robertson, *Girl Scout Council of the Nation's Capital* (Charleston, SC: Arcadia Publishing, 2013), 59–60.

18. Maryland–National Capital Park and Planning Commission, "Walker Mill Regional Park," http://www.pgparks.com/3236/Walker-Mill-Regional-Park.

19. Lynn Entwisle, email, October 14, 2019.

20. Prince George's County, Maryland, Register of Wills, estate no. RE-38479, liber 1839, p. 201 (copies provided August 15, 2019). Order blank for All Stone Memorials, Inc., Suitland, MD, April 8, 1991; price including sales tax was $1,258.75.

CHAPTER 9

1. In the early 1930s Ithaca's First National Bank moved from the location in figure 26 to a new, seven-story building a block away, at State and Tioga Streets.

2. The exceptions are Erie County, Pennsylvania, and Tioga County, New York.

3. Between 1861 and 1994 the maximum term of a US patent was seventeen years; it increased to twenty years in 1995. Mark Lemley, "An Empirical Study of the Twenty-year Patent Term," *AIPLA Quarterly Journal* 22 (1994): 369–424.

4. Tompkins County Clerk, Corporations Books, Laserfiche, 1930–1964, Book #7, 83 (accessed November 26, 2019). Original incorporation information is included in "Certificate of Change of Name of Rural Directories, Inc. to Brown-Engleson Publishing Corporation," recorded October 17, 1938, County Clerk control number BF104963-001.

5. "D. Boardman Lee Dies," *Ithaca Journal*, November 17, 1980, 3.

6. Engleson's four years of college was confirmed by Richard M. Engleson, 1940 Census, Williamson, Wayne County, New York, supervisor's district 36, enumeration district 59-55, sheet 2-B, Ancestry.com (November 27, 2019). Richard's sketchy biographical footprint did not name the college he had attended. Also see the Richard Michael Engleson family tree on Ancestry. The 1930 Census schedule reports his father, Michael O. den Englesen, as an immigrant from Holland, living in a $30,000 home, and working as a produce dealer. The 1920 Census schedule reported that the household included a servant, also from Holland, and that Engleson's father had also been a produce dealer a decade earlier. Richard had a brother, Leone, who was sixteen years older.

7. The entry, which identified him as "pres Rural Surveys Inc," indicated that he was living with his wife, Helen, in a house at 121 W. Court Street. *Manning's*

Ithaca (New York) Directory for the Year Beginning March 1939 (Schenectady, NY: H. A. Manning Co., 1939), 166. The couple was not listed in the directories for 1936, 1937, 1938, or 1940.

8. Engleson died on January 19, 1970, at age fifty-four, in Brownsville, Texas, apparently while on a business trip. His death certificate reported him as the "owner-operator" of a "farm implements" business and listed two causes of death, arteriosclerosis of coronary arteries with an "interval between onset and death" of fifteen years, and "acute myocardial infarction" (heart attack), which killed him in five minutes. Ancestery.com, Texas, Death Certificates, 1903–1982 [database online]; and "Regional Deaths," *Democrat and Chronicle* (Rochester, New York), January 21, 1970, 18.

9. "H. Stilwell Brown" [obituary], *Ithaca Journal*, November 9, 1987, 5.

10. "Elect Ithacans High Officers of the ABC," *Ithaca Journal*, May 18, 1931, 9.

11. Ithaca City Directory, various years.

12. "Firm to Issue Directory of Farmers," *Ithaca Journal*, October 15, 1937, 5.

13. "Certificate of Change of Name of Rural Directories, Inc. to Brown–Engleson Publishing Corporation."

14. "Certificate of Change of Name of Brown–Engleson Publishing Corporation to Rural Surveys, Inc.," filed February 2, 1939, County Clerk control number BF104150-001.

15. *Manning's Ithaca (New York) Directory for the Year Beginning March 1939* (Schenectady, NY: H. A. Manning Co., 1939), 135; and *Manning's Ithaca (New York) Directory for the Year Beginning March 1940* (Schenectady, NY: H. A. Manning Co., 1940), 130.

16. "New Company Aims to Map Rural Areas," *Ithaca Journal*, May 8, 1936, 1.

17. For a short overview of the mission of Ithaca Enterprises, see "Charter Given Group to Seek New Industries," *Ithaca Journal*, March 19, 1936, 3.

18. Onondaga County, NY. *Rural Index and Almanac, with Map* (Ithaca, NY: Rural Directories, 1936), 20.

19. Onondaga County, NY. *Rural Index and Almanac*, 21.

20. Onondaga County, NY. *Rural Index and Almanac*, 7 and 26–27.

21. Onondaga County, NY. *Rural Index and Almanac*, 13.

22. The additions were Carpenter and Foster. Onondaga County, NY. *Rural Index and Almanac, with Map*, 3.

23. *Compass System Map of Albany County, New York* (Ithaca, NY: Rural Surveys, 1939), 11 and 9.

24. "Joseph Short Heads College Alumni Program," *Ithaca Journal*, October 6, 1955, 5; "WHCU Official Retires," *Ithaca Journal*, May 11, 1971, 11; and "Joseph A. Short" [obituary], *Ithaca Journal*, May 29, 1993, 4.

25. *The Cayugan 1942–43* (yearbook), (Ithaca, NY: Ithaca College, 1943), 14; Census taker's manuscript for the 1940 Census, Ithaca, ward 5, Tompkins County, New York, supervisor's district 37, enumeration district 59-37, sheet 2-A, Ancestry.com (November 28, 2019); "Student Play Scheduled," *Ithaca Journal*, November 15, 1958, 2.

26. This paragraph refers to various pages in the Ithaca city directory for 1937 (pp. 134, 152, 212), 1938 (pp. 149, 204), 1939 (p. 153), 1940 (p. 149), and 1941 (p. 153).

27. "Albert MacWethy Chosen First Vicepresident, Carl Buchanan Secretary—B. A. Livingston Treasurer—Pratt Is Retiring Leader," *Ithaca Journal*, May 31, 1934, 9.

28. "Glenn Edward Bullock" [obituary], *Ithaca Journal*, January 25, 1983, 4.

29. "Roger Reid Dies; Insurance Agent," *Syracuse Post-Standard*, August 1, 1973, 10.

30. "Gilbert Hoffman Sidenberg, B.A., 1931," *Quarterly Record of Graduates of Yale University Deceased During the Year 1943–44*, series 41 (January 1, 1945), no. 1, 172–73.

31. "Ithaca Men Awarded Decorations," *Ithaca Journal*, July 2, 1945, 3.

32. "Food Firm Names Division Manager," *Ithaca Journal*, May 24, 1950, 5.

33. "Firm to Issue Directory to Farmers," *Ithaca Journal*, October 15, 1937, 5.

34. According to the 1938 Ithaca city directory, Ithaca Engraving Co. was at the corner of Seneca and Tioga Streets; Norton Printing was at 317 East State Street; and Rural Directories occupied rooms 701 and 702 on the top floor of the First National Bank Building, at the intersection of Tioga and East State Streets. *Manning's Ithaca (New York) Directory for the Year Beginning January 1938* (Schenectady, NY: H. A. Manning Co., 1938), 415, 417.

35. Taking Broome County as an example, representative titles were "1938 Farm Register of All Farms of Broome County, New York: For Merchants, Manufacturers, and Professional Men" as well as "1938 Special Classified Farm Register of All Farms of Wayne County, New York: For Merchants, Manufacturers, and Professional Men." An additional copyright was registered for the farm edition, titled *Rural Index Compass System Map and Localized Almanac, Broome County, New York*, which included the map in a pocket in the back of the directory. Library of Congress, Copyright Office, *Catalog of Copyright Entries*, Part 1: Books, Group 2, n.s., vol. 35, no. 5 (Washington, DC: Government Printing Office, 1938), 451, 564. By contrast, the map for Wayne County was copyrighted separately, as the *Compass System Map of Wayne County, New York*; see *Catalog of Copyright Entries*, Part 1: Books, Group 2, n.s., vol. 34, no. 11 (Washington, DC: Government Printing Office, 1937), 1158, 1168.

36. Quotations from the cover of the "1937–38 Special Classified Farm

Register of All Farms of Cortland County, New York: For Merchants, Manufacturers, and Professional Men" (Ithaca, NY: Rural Directories, 1937).

37. Madison County, New York, maintains a history website that includes key parts of 1937 *Special Classified Farm Register of All Farms of Madison County, New York: For Merchants, Manufacturers, and Professional Men* (Ithaca, NY: Rural Directories, 1937), including the list of each farmer's name, postal address, and Compass System map address. http://madisoncountynewyork.com/index.html.

38. For a concise (and intentionally humorous) history of telephone directories (and the only specific history I could find), see Ammon Shea, *The Phone Book: The Curious History of the Book That Everyone Uses But No One Reads* (New York: Perigee Books, 2010), esp. 53–66, 158–59. Telephone directories were well established as an advertising medium at the outset of the twentieth century.

39. The Census Bureau, which ran a separate survey of farmers, first inquired about a telephone in 1920, and published tabulations by state in 1920 and 1930, and by county from 1940 through 1964. For the *Census of Agriculture* data I relied on the digital historical archive maintained by Cornell University's Mann Library, online at http://agcensus.mannlib.cornell.edu/AgCensus/homepage.do.

40. John Brooks, *Telephone: The First Hundred Years* (New York: Harper and Row, 1975), 99–100, 115–17; and Claude S. Fischer, *America Calling: A Social History of the Telephone to 1940* (Berkeley and Los Angeles: University of California Press, 1992), esp. 99–107.

41. For a concise illustrated explanation of early telephone instruments and systems, see Herbert T. Wade, *Everyday Electricity: A Simple Introduction to Common Electric Phenomena* (Boston: Little, Brown, and Co., 1930), 126–53, esp. 129–34.

42. "Certificate of Increase in Capital Stock and Number of Shares of Rural Directories, Inc.," filed with New York Department of State on October 17, 1938, Tompkins County Clerk control number BF104700-001.

43. "Certificate of Reduction of Capital Stock and Number of Shares, Authorization of New Preferred Shares and Reclassification of Old Shares of Rural Surveys, Inc.," filed with New York Department of State on March 20, 1939, Tompkins County Clerk control number BF104152-001.

44. In 1940, Rural Surveys had shared 147–151 East State Street with the F. W. Woolworth Co. variety store, the annex of the N.Y. Electric & Gas Co., and William A. Church Co., at the rear. *Manning's Ithaca (New York) Directory for the Year Beginning March 1940* (Schenectady, NY: H. A. Manning Co., 1940), 349. The following edition showed the US Department of Agriculture, Soil Conservation

Service, as a new tenant, and listed "Rural Surveys Inc dir and map publishers" at 945 Cliff Street, identified elsewhere as the residence of H. Stilwell Brown, "slsm WHCU Radio Studio." *Manning's Ithaca (New York) Directory for the Year Beginning January 1941* (Schenectady, NY: H. A. Manning Co., 1941), 134, 272, 345. By 1942 the directory had dropped Rural Surveys altogether.

CHAPTER 10

1. Oddly, Wilmer Atkinson, who founded *The Farm Journal* in 1877, never mentioned the county directories in his 393-page posthumously published autobiography; see *Wilmer Atkinson: An Autobiography, Founder of the Farm Journal* (Philadelphia: Wilmer Atkinson Company and J. B. Lippincott, 1920). By the 1950s the *Journal* was the country's leading general interest agricultural periodical, and in the 1980s it began offering a broad range of data management and marketing services. D. A. Brown, "Agricultural Periodicals," *Library Trends*, 10 (January 1962): 405–13; and Benjamin J. Ventresca, Jr., "DataBase Marketing at Farm Journal, Inc.," *Journal of Direct Marketing* 5, no. 4 (Autumn 1991): 44–49.

2. A former Rand McNally salesman, Caleb Stillson Hammond left the company in 1900, apparently after a dispute about salary. Relocated in New York, Hammond was well situated to supply maps for book and encyclopedia publishers. Gerald A. Danzer, "Hammond Map Company (U.S.)," *HC6*, 574–77.

3. The *Catalog of Copyright Entries* reports scales of "1½ miles to 1 inch" and "3 miles to 2 inches" (both 1:95,040) and "2 miles to 1 inch" (1:126,720) for various *Farm Journal* maps, which publisher Wilmer Atkinson Co. typically registered separately from their accompanying directory.

4. WorldCat.org lists *Farm Journal* rural directories for Broome (published in 1917), Cortland (1917), Erie (1917), Genesee (1917), Livingston (1917), Madison (1917), Monroe (1918), Onondaga (1917), Orleans (1918), Steuben (1917), and Tioga (1917) Counties. Of these eleven counties, Plato published directories for only Broome (1922), Cortland (1921), Genesee (1929) Monroe (1922), Onondaga (1927), and Tioga (1922) Counties.

5. Edward Higbee, *American Agriculture: Geography, Resources, Conservation* (New York: John Wiley & Sons, 1958), 274, 296.

6. Steve Craig, "The Farmer's Friend: Radio Comes to Rural America, 1920–1927," *Journal of Radio Studies* 8 (2001): 330–46; Robert L. Hilliard, "Symposium Introduction: Farm and Rural Radio—Some Beginnings and Models," *Journal of Radio Studies* 8 (2001): 321–29; and Joann Vanek, "Work, and Family Work, Leisure, and Family Roles: Farm Households in the United States, 1920–1955," *Journal of Family History* 5 (1980): 422–31, esp. 427–28.

7. Postal zone numbers inserted between the city and state names—for

example, Baltimore 7, Maryland—were introduced at 124 of the country's largest cities to address the loss of many experienced postal clerks to military service during World War II. When mail arrived from out of town, clerks quickly sorted letters into bins headed to specific local post offices. US Postal Service, *The United States Postal Service: An American History, 1775–2006* (Washington, DC, 2007), 33. The five-digit ZIP (Zone Improvement Plan) Code was introduced nationwide in 1963, and the preceding zone-number address became Baltimore, MD 21207. *The United States Postal Service: An American History*, 43. Bulk mailers, who were required to presort by ZIP Code in 1967, readily accepted the ZIP+4 refinement, which tacked on a hyphen and four additional digits in 1983. *The United States Postal Service: An American History*, 43. Technology's role in expediting mail delivery is readily apparent in optical character recognition (OCR) equipment, improved incrementally since its initial introduction in the mid-1960s; in database software for converting electronic house-number addresses into nine-digit ZIP Codes; and in high-speed electromechanical systems that read barcodes and cope with letters and packets of different shapes and sizes. *The United States Postal Service: An American History*, 43–49.

8. Mark Monmonier, *Connections and Content: Reflections on Networks and the History of Cartography* (Redlands, CA: Esri Press, 2019), 171–75.

9. Or every fifty-three feet, to be more precise.

10. Margaret Corwin, who reviewed strategies for house-numbering for the American Society of Planning Officials, categorized Denman's approach as a "variant" of the "uniform" or "equal interval numbering" approach. Denman's unique contribution was the use of a car's odometer to assign numbers based on "distance of road covered." Margaret A. Corwin, *Street-Naming and Property-Numbering Systems*, Planning Advisory Service report no. 332 (Chicago: American Planning Association; American Society of Planning Officials, 1978), 14.

11. Ralph H. Denman, *Rural Planning and House Numbering* (Ithaca, NY: R. H. Denman, 1968), "Installation Diagram," 12A.

12. Denman, *Rural Planning and House Numbering*, 24.

13. Denman, *Rural Planning and House Numbering*, 27.

14. Denman, *Rural Planning and House Numbering*, 2. According to a note preceding the short book's preface, Denman was an "Agricultural Engineer, Rural Consultant and former Associate Professor of Rural Engineering at the University of Massachusetts [who] has worked with many rural towns."

15. Val Noronha, "Wayfinding and Travel Maps: In-Vehicle Navigation System," *HC6*, 1716–22.